D1080453

Universal Mobile Telecommunications System Security

UMTS Security

Valtteri Niemi and Kaisa Nyberg

Nokia Research Center, Finland

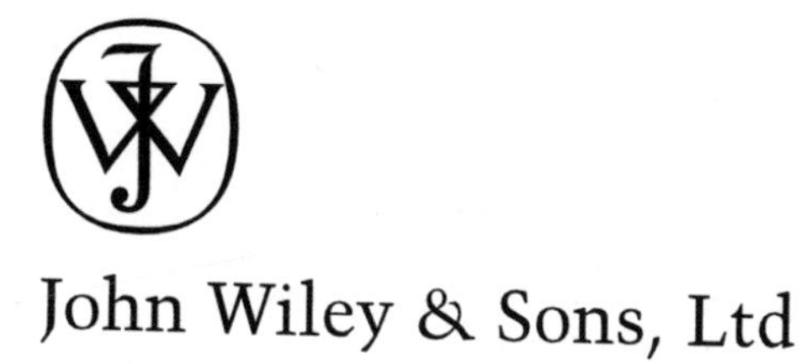

John Wiley & Sons, Ltd

Email (for orders and customer service enquiries): cs-books@wiley.co.uk
Visit our Home Page on www.wileyeurope.com or www.wiley.com

Other Wiley Editorial Offices

John Wiley & Sons, Inc., 111 River Street, Hoboken, NJ 07030, USA

Jossey-Bass, 989 Market Street, San Francisco, CA 94103-1741, USA

Wiley-VCH Verlag GmbH, Boschstr. 12, D-69469 Weinheim, Germany

John Wiley & Sons Australia Ltd, 33 Park Road, Milton, Queensland 4064, Australia

John Wiley & Sons (Asia) Pte Ltd, 2 Clementi Loop #02-01, Jin Xing Distripark, Singapore 129809

John Wiley & Sons Canada Ltd, 22 Worcester Road, Etobicoke, Ontario, Canada M9W 1L1

Wiley also publishes its books in a variety of electronic formats. Some content that appears in print may not be available in electronic books.

Library of Congress Cataloging-in-Publication Data

Niemi, Valterri.
UMTS security / by Valterri Niemi and Kaisa Nyberg.
p. cm.
ISBN 0-470-84794-8
1. Computer security. 2. Computer networks—Security measures.
3. Mobile communication systems—Security measures. 4. Wireless communication systems—Security measures. 5. Global system for mobile communications—Security measures. I. Nyberg, Kaisa, 1948– II. Title.
QA76.9.A25 N54 2003
005.8—dc22 2003022090

British Library Cataloguing in Publication Data

A catalogue record for this book is available from the British Library

ISBN 0-470-85314-X

Project management by Originator, Gt Yarmouth, Norfolk (typeset in 10/13pt Times)
Printed and bound in Great Britain by Antony Rowe, Chippenham, Wiltshire
This book is printed on acid-free paper responsibly manufactured from sustainable forestry in which at least two trees are planted for each one used for paper production.

Contents

Preface

Physical communication channels in wireless technology are inherently insecure. As a wire only has two ends, the wired communication technology has a means of supporting integrity of communication, whereas in wireless communication dedicated technology is needed—even to control a basic point-to-point connection. This technology of communication integrity, which forms an essential part of any modern wireless communications system, is called security technology.

This book is a description of the security solutions specified for the *Universal Mobile Telecommunication System* (UMTS). It gives a comprehensive presentation of UMTS security specifications and explains the role of the security functionality in UMTS. In the first place, this book is aimed at presenting the UMTS security system in its totality to planners, builders and implementers of UMTS networks. It also gives a unified treatment of the security services provided by UMTS that we expect will be invaluable to developers and analysts of application-oriented security services that make use of UMTS communication networks. This book could also serve as a textbook for an advanced university course in modern communication security technology.

To achieve global relevance a communication system requires standardization. Standardization guarantees that entities in the system are able to communicate with each other even when they are controlled by different mobile network operators or are manufactured by different vendors. However, it is important to have a non-standardized area that allows differentiation between operators and manufacturers. For instance, the internal structure of network entities belongs to the non-standardized area.

Security is visible in both the standard and the non-standard specifications of UMTS. As an example, communication between the mobile phone and the radio network is protected by encrypting the messages. It is important that strict standardization applies to both how encryption is carried out and which encryption keys are used, otherwise the receiving end could not reverse the operation and recover the original content of the message. On the other hand, both communicating parties have to store the encryption keys in such a way that no outsider can get access to them. It is important that this is done, but we do not have to standardize how it is done. The emphasis of our book is on the standardized features of UMTS security but not at the expense of other aspects.

The book is split into two parts. The first describes the security architecture and security functionality of UMTS, while the specification and analysis of cryptographic algorithms is presented in the second part. By breaking the book down in this way we hope to offer a specialized treatment of the two different areas and methodologies that comprise the UMTS security system. While the parts can be read independently, we feel by including them in one book that the reader has the chance to become familiar with the delicate interplay between security and cryptography.

This book presents the results of the extensive, demanding and strenuous work of security expert teams and individuals who together created the specifications of UMTS security. It draws largely on collaboration and discussions with the world's foremost specialists all of whom participated in this work. In particular, we want to express our thanks to the members of 3GPP SA3, ETSI SAGE and the 3GPP algorithms task forces and, of course, our colleagues at NOKIA.[1]

Finally we would like to thank the publisher of the book and the editing team whose splendid work transformed our typescript into a coherent book.

[1] This book represents the views and opinions of the authors and does not necessarily represent the views of their employer or any standards organization. In particular, no part of the material contained in this book should be used as a 3GPP or IETF specification.

Part I

Security Architecture for UMTS

1

Introduction to Security and to UMTS

1.1 Security in Telecommunications

The wireline communication network has certain inherent security properties: the network operator has basically installed the whole network including lines all the way to the socket in the wall and the user simply plugs a terminal wire into the socket. This means that the operator has some assurance that the configuration of the network remains fixed during the operation, making it easy to determine which subscribers to charge for each call.

It is also clear that no accidental bypasser can eavesdrop conversation during a fixed line phone call. Indeed, active wiretapping has to occur before eavesdropping is possible or the switch has to be broken into. From the average user's point of view, the probability of these threats seems remote enough for the user to be reasonably confident that no foul play is going on during a fixed line phone call.

On the other hand, wiretapping is not a technically-demanding task. If someone is really determined to listen to a certain user's communication, it certainly can be done. Moreover, it is not much more difficult to place phone calls on the "attacked" line. There are several reasons why these things do not happen often in practice: the fact that the incentive to carry out an active attack is low compared with the cost of the work needed and, of course, the whole activity is illegal.

A private detective could in principle be tempted to actively wiretap on behalf of a jealous husband or wife, and somebody who needs to communicate a lot over long distances may be temped to try to put the cost on some big company's phone bill. However, it is difficult to run these activities for a long time without someone noticing, especially if the activity is large scale. If the thief starts to sell stolen phone calls, the victim inevitably notices that something is wrong and the "attacker" gets caught sooner or later.

The situation changes considerably when a cordless phone is used. Radio communication can be eavesdropped from short distances and the risk of getting caught is small compared with a wiretapping attack. However, hijacking calls is still a risky activity given that it has to be done from the immediate proximity.

In the case of a cellular network the eavesdropping attack can be carried out over a relatively large area. The active stealing of calls is fairly easy, in principle, given the fact that the network has no real control over user movements. Indeed, in first-generation mobile networks, where analogue techniques were in use, listening to other people's phone calls became a popular pastime. Also, in some of the first analogue systems, charging was simply based on the user's own announcement of the number dialled.

The second-generation (2G) mobile networks use digital technology instead of analogue signals. This offers a new tool that can be used to counter security problems created by the introduction of wireless networks. Indeed, it is possible to digitally manipulate the signals (e.g., error-correcting codes can be used to reduce the effects of disturbances in the radio channel). In the security area, cryptographic methods can be used: for confidentiality of calls encryption can be used and to prevent calls being stolen cryptographic authentication mechanisms can be utilized. The Global System for Mobile Communications (GSM) is the largest 2G mobile network and its security system was the starting point of the development of security features for subsequent generations.

1.1.1 General security principles

Security is an abstract concept: it is not easy to define but people tend to understand quite well what kind of issues are dealt with when speaking about it. Protection methods to counter criminal activity lie at the very core of security. There is a clear distinction between security on the one hand and fault tolerance and robustness on the other.

In telecommunication systems, many aspects of security are relevant. There is a clear differentiation between *physical* security and *information* security. The first consists of locked rooms, safes, guards, etc. All these are needed when operating a large-scale telecommunication network. Another element of physical security is tamper resistance. Smart cards play a major role in the system we describe in this book mainly because they are tamper-resistant. Sometimes the visibility of tamper evidence is sufficient protection against physical intrusion against network elements. If tampering can be detected quickly enough, corrupt elements can be cut out of the network before too much damage is caused.

Biometric protection mechanisms can be seen as examples of methods that are intermediate between physical security and information security (e.g., fingerprint checking assumes both sophisticated measurement instruments and sophisticated information system to support the use of these instruments as access control devices). In the future, biometrics is likely to become an increasingly important component of security systems.

1.1.1.1 Communication security

In this book we concentrate on those aspects of telecommunication security that belong to the broad category of information security. More specifically, we deal with *communication* security. Another division of information security can be derived from a conceptual point of view. The following areas can be studied fairly independently of each other:

1. *System-level* security. Here the leading principle is "the system is as strong as its weakest link." Attackers always try to find a point weak enough to be broken and once inside the system the next steps are often easier to do.
2. *Application* security. For instance, banking applications over the Internet typically use security mechanisms that are tailored to meet their specific requirements.
3. *Protocol-level* security. This is about how communicating parties can achieve security goals by executing well-defined communication steps in a predefined order.
4. *Operating system security*. The behaviour of all elements in a network, including mobile terminals, depends on the correct functionality of the operating system that controls them.
5. Security *primitives*. These are the basic building blocks on top of which all protection mechanisms are built. Typical examples are cryptographic algorithms, but also items like protected memory space can be seen as security primitives.

In this book we put the main emphasis on areas 1, 3 and 5 while the rest are touched on only briefly.

When dealing with the design of practical security systems it is important to remember that there are also tight constraints. Cost of implementing protection mechanisms must be balanced against the risks that are reduced by such mechanisms. In addition, the usability of the system should not suffer because of the security mechanisms. Here both user aspects and the general performance of the system are relevant. It is easy to build an extremely secure system if there are no limitations on the cost of building or operating the system and all users are assumed to follow cumbersome procedures (without making errors) for the sake of security. In military systems, trade-offs between security, cost and usability are of course done on a totally different basis than in public or general purpose communication systems.

1.1.1.2 Design of a secure system

The design process of a security system contains the following phases:

- *Threat analysis*. The intention here is to list all possible threats against the system. At this phase there is no need to find out what kinds of actions and devices are needed to carry out an attack to realize the threat.
- *Risk analysis*. In this phase the weight of each threat is measured quantitatively or, at least, in relation to other threats. This requires estimating both the complexity of the various attacks and the potential damage caused by them.
- *Requirements capture*. Based on earlier phases, we now decide what kind of protection is needed for the system. The requirements are easiest to define in a context of a *trust model* that has to be defined first.
- *Design phase*. In this phase the actual protection mechanisms are designed in order to meet the requirements. Existing building blocks (e.g., security protocols or primitives) are identified, possibly new mechanisms are created and a security architecture is built. Here constraints have to be taken into account, and it is possible that all requirements cannot be met, implying a revisit to earlier phases, especially to analysis.
- *Security analysis*. It is important to carry out an evaluation of the results independently of the previous phase. Straightforward verification tools can usually be utilized only for limited parts of a security analysis, while often "holes" in the security system can only be revealed by methods that are sufficiently creative.
- *Reaction phase*. While planning of the system management and operation can be seen as part of the mechanism design phase, reaction to all unexpected security breaches cannot be planned beforehand. In the reaction phase it is vital that the original design of the system is flexible enough to allow enhancements. To reduce the difficulties in the reaction phase, some safety margin should be built into the mechanisms that are designed. These margins tend to be useful in cases where new attack methodologies and tools are developed faster than expected.

We have listed here only those phases that are involved in the design process. Naturally, *implementation* and *testing* phases play major parts of the total effort of building a secure system.

One additional constraint that affects several phases is due to the fact that the security system is often part of a much larger system that is designed at the same time. This is the case for communication networks also and has been a factor in the specification work of third-generation (3G) mobile networks. In practice, this implies that some iterations are needed because the general system architecture and requirements are changing while the security design work is ongoing. However, it is important that the security for the system is designed at the same time as the system itself is designed. Trying to add security to an existing complete system typically leads to impractical and inefficient solutions.

1.1.2 GSM security

The goal of the security design for the GSM was clear: the security has to be as good as that of wireline systems and, at the same time, the mechanisms introduced were not allowed to reduce the usability of the system.

The goals were successfully reached. It can be argued that GSM has even better security than wireline systems. Nevertheless, it is also clear that the security of GSM could be even better. As time passes, attack methods and equipment evolve, but protection methods should also become increasingly better over time. Some enhancements on GSM security have been made over the years, but the basic structures have remained the same. It is always difficult to introduce radical changes into a working system. This illustrates one of key principles that should be followed when protection mechanisms for future systems are developed: they should not only provide adequate protection against contemporary attack techniques but there should also be an additional safety margin. It is of course difficult to predict what kinds of tools attackers may use in the future, but the following guideline should at least be taken into account: all theoretical attack scenarios tend to become practical attacks sooner or later.

The most important security features in the GSM system are:

- authentication of the user;
- encryption of communication in the radio interface;
- use of temporary identities.

As GSM and other 2G systems become increasingly successful, the usefulness of these basic security features also becomes increasingly evident. Naturally, it was paramount in the specification work of the UMTS security to carry these features over to the new system.

The success of GSM also emphasized the limitations of its security. Popular technologies are tempting for fraudsters. The properties of GSM that have been most criticised on the security front are the following:

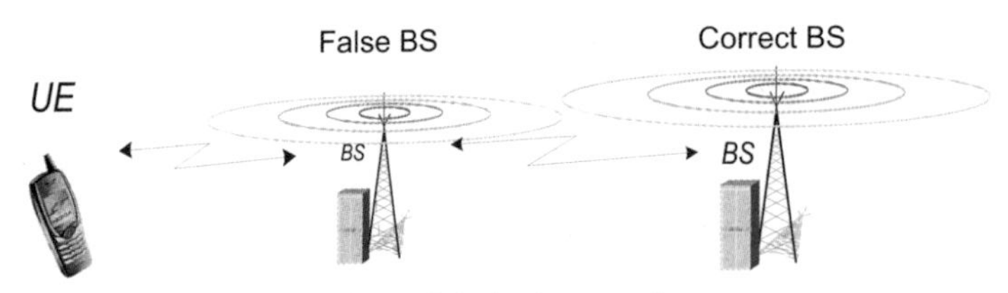

Figure 1.1 Active attack
UE = User Equipment, BS = Base Station

- active attacks on the network are possible, in principle, by somebody who has the requisite equipment to masquerade as a legitimate network element and/or legitimate user terminal (see Figure 1.1 for an example scenario);
- sensitive control data (e.g., *keys* used for radio interface ciphering) are sent between different networks without cryptographic protection;
- some essential parts of the security architecture are kept secret (e.g., the cryptographic algorithms), creating a lack of trust in them in the long run because they are not available for analysis by most recently developed methods and, in any case, such secrets tend to surface eventually.

Keys used for radio interface ciphering eventually become vulnerable to massive *brute force* attacks where somebody tries all possible keys until one makes a match.

These limitations were retained in the GSM system on purpose. The threat posed by them was considered small in comparison with the added cost of trying to circumvent them. However, as technology advances, attackers gain access to better tools. That is why the outcome of a similar comparison between cost and security led to a different conclusion in the case of 3G mobile networks.

In the next sections we take a brief look at the most important GSM security features.

1.1.2.1 Authentication of the user in GSM

There exists a permanent secret key Ki for each user *i*. As depicted in Figure 1.2 this key is stored in two locations:

- in the user's Subscriber Identity Module (SIM) card;
- at the Authentication Centre (AuC).

The key Ki never leaves either of these two locations. *Authentication* of the user is based on the idea of checking whether the user has access to Ki. This can be achieved by *challenging* the user to do a computation that can only be done with the key Ki. A

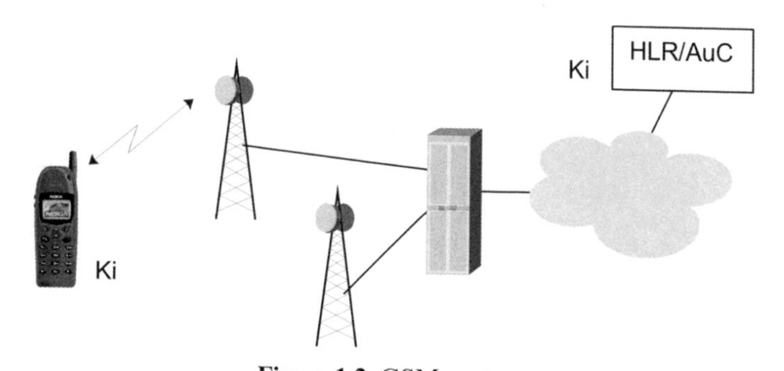

Figure 1.2 GSM system
GSM = Global System for Mobile communications; HLR = Home Location Register; AuC = Authentication Centre

random 128-bit string RAND (random number) is sent to the mobile phone. The phone transfers the parameter to the SIM card (inside the phone). In SIM, there is a *one-way function*, denoted by A3, that takes two inputs: Ki and RAND. The output is a 32-bit response value SRES (signed response) that is sent back to the network where the correctness of the response is checked.

Similarly, a temporary *session key* Kc is generated as an output of another one-way function A8 that takes the same input parameters Ki and RAND. This key is used to encrypt phone calls on the radio interface. The serving network has no knowledge of the master key Ki and, therefore, it cannot handle all of the security alone. Instead, other relevant parameters (i.e., the so-called *authentication triplet*—RAND, SRES, Kc) are sent to the serving network element MSC/VLR (Mobile Switching Centre/Visitor Location Register), or SGSN (Serving GPRS Support Node) in the case of GPRS (General Packet Radio Service) from the AuC. The process of identification, authentication and cipher key generation is depicted in Figure 1.3.

1.1.2.2 GSM ciphering

During authentication a secret session key Kc is established. With this key all calls are encrypted between the phone and the base station until the next authentication occurs.

The encryption algorithm is called A5. It is standardized but the specification is still confidential; it is managed by the GSM Association and delivered under specific licence to vendors that produce GSM equipment, either terminals or base stations. Figure 1.4 describes the high-level structure of the GSM ciphering algorithm A5.

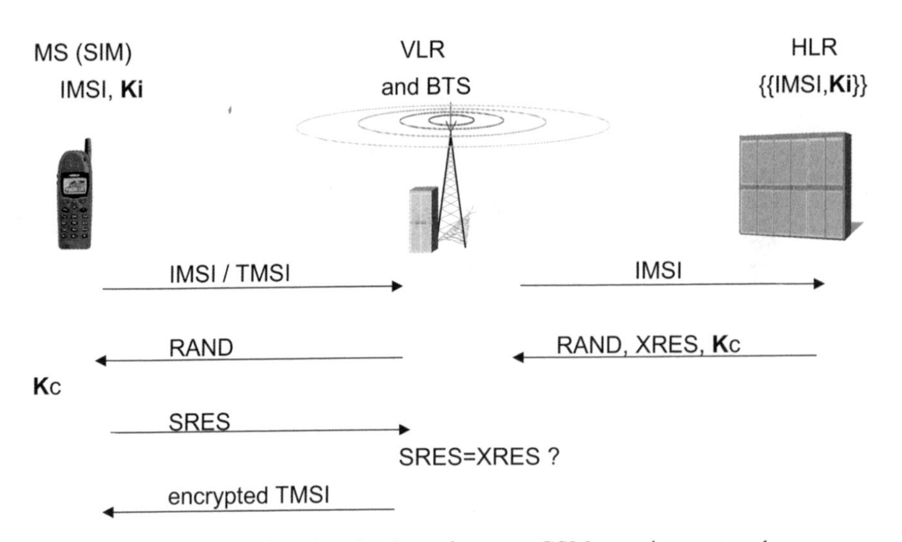

Figure 1.3 Identification and authentication of a user: GSM security protocol
GSM = Global System for Mobile; MS = Mobile Station; SIM = Subscriber Identity Module; IMSI = International Mobile Subscriber Identity; VLR = Visitor Location Register; BTS = Base Transceiver Station; HLR = Home Location Register; TMSI = Temporary Mobile Subscriber Identity; RAND = random number; XRES = expected response; SRES = signed response

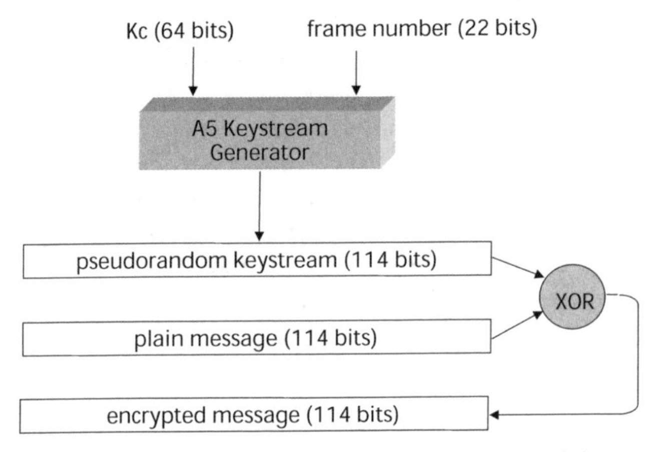

Figure 1.4 GSM encryption: structure of A5 stream cipher

1.1.2.3 GPRS ciphering

In GPRS, radio interface ciphering by the algorithm A5 is replaced by another stream cipher, called GEA (GPRS Encryption Algorithm). This change was made because the termination point of encryption was moved deeper in to the network, from the base station to SGSN. Naturally, this also implies that the ciphering function is now applied at a higher layer. In (circuit-switched) GSM, encryption is done at the physical layer, while in GPRS encryption is done at layer 3, more specifically, at the Logical Link Control (LLC) layer.

1.1.2.4 User identity confidentiality

The permanent identity of the user—International Mobile Subscriber Identity (IMSI)—is protected in GSM against eavesdroppers by restricting the number of occasions where it has to be used. Instead of IMSI, a Temporary Mobile Subscriber Identity (TMSI) is normally used for identification of the user. The TMSI is changed every time it has been used and the new TMSI is always transmitted to the user over the encrypted channel. A similar mechanism is used also in UMTS and is described in more detail in Section 2.1.2.

In GPRS, a separate temporary identity—Packet TMSI (P-TMSI)—is used. It is allocated independently of the TMSI by the packet core network element SGSN.

1.2 The Background to 3G

The roots of mobile communications are in military applications and date back to the early phases of radio technology. The concept of a cellular network to serve a large number of mobile users was taken into commercial use much later, at the beginning of the 1980s, in the form of the Advanced Mobile Phone System (AMPS) in the USA and in the form of the Nordic Mobile Telephone (NMT) system in Northern Europe. The first generation of cellular systems was based on analogue technologies. Typically, Frequency Modulation (FM) radio was used and simultaneous access by many users in the same cell was provided by the Frequency Division Multiple Access (FDMA) technique. Handovers between different cells were already possible to ensure continuous connections while the user was moving. This was essential because the dominant service type involved phone calls from cars. Limited roaming between networks was also included in the first-generation mobile networks.

At the beginning of the 1990s, 2G mobile systems were introduced. The most successful of them has been GSM, which had more than 800 million users worldwide at the beginning of 2003. The leading 2G technology has been TDMA (Time Division Multiple Access) in the USA and PDC (Personal Digital Cellular) in Japan. The most important new feature in 2G was the introduction of digital information transmission in the radio interface between the mobile phone and the base station. In all these systems, the multiple access technology is TDMA.

The most immediate advantages of 2G over its predecessor were increased capacity of the network (due to more-effective use of radio resources), better speech quality (due to digital-coding techniques) and the possibility for communicating data much more easily. Also, it was now possible to enhance the security of the system significantly.

At the same time as 2G systems were launched, it became clear that there was also a next step to be taken at some point. The work to design the 3G system was initiated in organizations like European Posts and Telecommunications Conference (CEPT) and the UMTS Forum, and later the European Telecommunications Institute (ETSI) began to develop the work further. One of the leading ideas for 3G was to ensure fully global roaming: to make it possible for the user to use mobile system services all over the world. In the global International Telecommunication Union (ITU), this goal was proposed for the "IMT-2000" (International Mobile Telecommunications) standard.

1.3 The 3G Partnership Project (3GPP)

The phenomenal success of GSM had a twofold effect on the development of the new generation system. From a positive point of view, the success of mobile communication technologies made it easier to find resources for subsequent research and development. It was soon clear that GSM would not be the end point of the road in mobile technology.

From a negative point of view, there seemed to be no immediate need for a new system since GSM had proven to be such an effective system. Thus, for several years development of UMTS was done on a theoretical basis only. On the security side, a great deal of effort was put in (e.g., development of new authentication mechanisms). Many state-of-the-art cryptographic techniques were proposed for UMTS security; however, it was not possible to decide between different proposals and options as the constraints imposed by the system architecture and, for instance, the radio technologies had still not been clarified.

Toward the end of the century it became evident in Japan that the 2G system PDC (Personal (or Pacific) Digital Cellular) was not going to provide a good enough service for the huge regional market in the near future. The radio capacity of PDC was simply eaten up by demand. Simultaneously with ETSI work, Japanese standard organizations ARIB (Association of Radio Industries and Businesses) and TTC (Telecommunications Technology Council) were creating detailed specifications for 3G technology, especially the radio network part. In 1998, five standards organizations decided to combine their efforts to accelerate the work and guarantee global interoperability. The organizations, ETSI from Europe, ARIB and TTC from Japan, T1 from North America, and TTA (Telecommunications Technology Association) from South Korea formed the 3GPP. Soon afterward, a sixth partner from China joined the project. The current Chinese partner in 3GPP is CCSA (China Communications Standards Association).

The crucial technical decision was to base radio access technology on Wideband Code Division Multiple Access (WCDMA). On the core infrastructure side, it was decided to reuse the GSM model and develop the 3GPP core network from GSM. Another decision was to include the concept of SIM in an evolved form in the new system. This latter decision was especially important for security purposes and was the decision about core network technology that led to another decision that placed security specification in the hands of the 3GPP, taking ETSI draft specifications as baseline documents.

The dream of getting all 3G development work under one project did not, however, come true. In the USA, much work had been done on a system called CDMA2000, which evolved from North American 2G legacy systems. Driven by the TIA (Telecommunications Industry Association), another project was started, called 3GPP2. At the same time in the ITU the original target of a single IMT-2000 standard was replaced by a family of 3G standards.

The 3GPP set an ambitious goal of creating technical specifications for the new system by 1999. Co-operation between the partners quickly began to work very well, and a large number of specifications were in a reasonably stable state at the end of 1999. In March 2000, the so-called "Release 1999" of the 3GPP specification set was declared "frozen". Even after that date, many corrections have been needed in most specifications, a process that is unavoidable in a project of this scale. After Release 1999, 3GPP started to create the next release, called Release 4, which was frozen in June 2001. Subsequently, the Release 5 specification set was frozen. However, it was believed that a cycle of one year was too short to produce a significant number of added features to warrant a new release and so 3GPP stopped linking releases to calendar years.

The specification work in 3GPP follows a three-stage model:

- stage 1 specifications define the requirements for new services;
- in stage 2 specifications an architecture is created that meets these requirements, including description of functional entities and information flows between them;
- in stage 3 specifications the functional entities are mapped into physical entities and bit-level descriptions of protocols between the entities are defined.

In addition to these specifications, there is also a set of test specifications that are typically completed some time later (because they are not needed until later).

In Table 1.1 we show how 3G specifications are divided into different series. For the purposes of this book, the 33 and 35 series are the most important.

Table 1.2 contains all the specifications and reports from Release 1999 onward that are (at the time of writing) under the responsibility of the 3GPP security working group 3GPP TSG SA WG3. Note that for many of the specifications Release 6 is going to be created later.

Table 1.1 Specification numbers

Number of series	Subject of series
21	Requirements
22	Service aspects
23	Technical realization
24	Signalling protocols (UE network)
25	Radio aspects
26	Codecs
27	Data
28	Signalling protocols (radio network–core network)
29	Signalling protocols (intrafixed network)
30	Program management
31	USIM
32	O&M
33	Security aspects
34	Test specifications
35	Security algorithms

UE = User Equipment; USIM = Universal Subscriber Identity Module; O&M = Operation and Maintenance

1.4 3GPP Network Architecture

In this section we give a brief overview of the 3GPP network architecture. A more thorough description can be found in [65] and a wider presentation of cellular networks can be found [96].

A simplified picture of the most important elements in the 3GPP Release 1999 system is given in Figure 1.5.

The network model consists of three main parts, all of which are visible in Figure 1.5. From the user point of view, the most visible part is the *terminal*, which is also called the User Equipment (UE). The terminal has a radio connection to the local (Radio) Access Network (RAN), which in turn is connected to the Core Network (CN). Among other things, the CN takes care of the global aspects of the system.

The CN contains two main *domains*: the Packet Switched (PS) domain and the Circuit Switched (CS) domain. The former is an evolution of the GPRS domain, and the most important network elements in the PS domain are the SGSN and the Gateway GPRS Support Node (GGSN). The CS domain is an evolution from the traditional, CS GSM network with the MSC as the most important component.

In addition to the various network elements, the architecture defines *interfaces* between these elements. We also need *protocols* to define how different elements are able to communicate over the interfaces. Protocols involving UE are grouped into two main *strata*: the *access stratum* contains protocols that are run between the UE and the access network and the *non-access stratum* contains protocols between the UE and the CN. In addition, there are lots of protocols that are run between different network elements.

The CN in the general network model can also be divided into two parts: the *home* network or home entertainment (HE) and the *serving* network (SN). The home network contains all the static information about the *subscriptions* of the users, including the static security information. The serving network locally handles communication to the UE (via the access network). In the event the user is *roaming*, the home and the serving network are controlled by different *mobile network operators*.

1.4.1 Elements in the architecture

Let us have a closer look at the most important elements of the 3GPP architecture. Looking first at the terminal, the UE consists of two parts: the Mobile Equipment (ME) and the Universal Subscriber Identity Module (USIM). The ME is typically the mobile phone that contains the radio functionality and all the protocols that are needed for communication with the network. It also contains the user interface (e.g., display and keypad). The USIM is contained in a smart card, which is placed inside the ME. The USIM contains all the operator-dependent data about the subscriber, including the permanent security information.

There are two types of radio access networks in the 3GPP system. The new, revolutionary access network is called the UMTS Terrestrial Radio Access Network (UTRAN) and is based on the W-CDMA technology. The bulk of the specification work in the 3GPP Release 1999 was devoted to the development of UTRAN functionalities. In the subsequent 3GPP Release 4, another RAN type was introduced to the system. This alternative is more evolutionary and is called GSM/EDGE Radio Access Network (GERAN). It is based on a new modulation technology that is likely to allow data rates on the GSM network to triple. On the other hand, certain key features of UTRAN have also been introduced in GERAN. These key features also include several security features.

UTRAN contains two types of elements. The Base Station (BS) is the termination point of the radio interface (Uu) on the network side and is called Node B in the 3GPP architecture. We mainly use the term "base station" in this book instead of the more technical term "Node B". The BS is connected (over the Iub interface) to the controlling unit of UTRAN (i.e., to the Radio Network Controller (RNC)). The interface between the RNC and the CN is called the Iu interface.

As a new feature (compared with the GSM BS subsystem), it is possible in UTRAN to connect RNCs directly to each other over the Iur interface. This makes it possible to manage the radio resources and mobility of the users more effectively. Also, the UE may be connected to the network via several RNCs

Table 1.2 Security specifications and reports

			R1999	R4	R5	R6
TS	21.133	3G security; security threats and requirements	✓	✓		
TS	22.022	Personalization of ME; mobile functionality specifications	✓	✓	✓	
TS	22.031	FIGS; service description; Stage 1	✓	✓	✓	
TS	22.032	IST; service description; Stage 1	✓	✓	✓	
TS	23.031	FIGS; service description; Stage 2	✓	✓	✓	
TS	23.035	IST; Stage 2	✓	✓	✓	
TS	33.102	3G security; security architecture	✓	✓	✓	
TS	33.103	3G security; integration guidelines	✓	✓		
TS	33.105	Cryptographic algorithm requirements	✓	✓		
TS	33.106	Lawful interception requirements	✓	✓	✓	
TS	33.107	3G security; lawful interception architecture and functions	✓	✓	✓	
TS	33.120	Security objectives and principles	✓	✓		
TR	33.901	Criteria for cryptographic algorithm design process	✓	✓		
TR	33.902	Formal analysis of the 3G authentication protocol	✓	✓		
TR	33.908	3G security; general report on the design, specification and evaluation of 3GPP standard confidentiality and integrity algorithms	✓	✓		
TS	35.201	Specification of the 3GPP confidentiality and integrity algorithms; Document 1: f8 and f9 specifications	✓	✓	✓	
TS	35.202	Specification of the 3GPP confidentiality and integrity algorithms; Document 2: Kasumi algorithm specification	✓	✓	✓	
TS	35.203	Specification of the 3GPP confidentiality and integrity algorithms; Document 3: implementors' test data	✓	✓	✓	
TS	35.204	Specification of the 3GPP confidentiality and integrity algorithms; Document 4: design conformance test data	✓	✓	✓	
TS	33.200	3G security; NDS; MAP application layer security		✓	✓	
TR	33.903	Access security for IP-based services		✓	✓	
TR	33.909	3G security; report on the design and evaluation of the MILENAGE algorithm set; Deliverable 5: an example algorithm for the 3GPP authentication and key generation functions		✓ ✓		
TS	35.205	3G security; specification of the MILENAGE algorithm set: an example algorithm set for the 3GPP authentication and key generation functions f1, f1*, f2, f3, f4, f5 and f5*; Document 1: general		✓	✓	
TS	35.206	3G security; specification of the MILENAGE algorithm set: an example algorithm set for the 3GPP authentication and key generation functions f1, f1*, f2, f3, f4, f5 and f5*; Document 2: algorithm specification		✓	✓	
TS	35.207	3G security; specification of the MILENAGE algorithm set: an example algorithm set for the 3GPP authentication and key generation functions f1, f1*, f2, f3, f4, f5 and f5*; Document 3: implementors' test data		✓	✓	
TS	35.208	3G security; specification of the MILENAGE algorithm set: an example algorithm set for the 3GPP authentication and key generation functions f1, f1*, f2, f3, f4, f5 and f5*; Document 4: design conformance test data		✓	✓	
TR	35.909	3G security; specification of the MILENAGE algorithm set: an example algorithm set for the 3GPP authentication and key generation functions f1, f1*, f2, f3, f4, f5 and f5*; Document 5: summary and results of design and evaluation		✓	✓	

			R1999	R4	R5	R6
TR	41.031	FIGS; service requirements; Stage 0		✓	✓	
TR	41.033	Lawful interception requirements for GSM		✓	✓	
TS	41.061	GPRS; GPRS ciphering algorithm requirements		✓		
TS	42.009	Security aspects		✓		
TS	42.033	Lawful interception; Stage 1		✓	✓	
TS	43.020	Security-related network functions		✓	✓	
TS	43.033	Lawful interception; Stage 2		✓	✓	
TS	33.201	Access domain security			✓	
TS	33.203	3G security; access security for IP-based services			✓	
TS	33.210	3G security; NDS; IP network layer security			✓	
TS	33.900	Guide to 3G security			✓	
TS	33.108	3G security; handover interface for lawful interception			✓	✓
TR	33.810	3G security; NDS/AF; feasibility study to support NDS/IP evolution				✓
TS	55.205	Specification of the GSM MILENAGE algorithms: an example algorithm set for the GSM authentication and key generation functions A3 and A8				✓
TS	55.216	Specification of the A5/3 encryption algorithms for GSM and EDGE, and the GEA3 encryption algorithm for GPRS; Document 1: A5/3 and GEA3 specification				✓
TS	55.217	Specification of the A5/3 encryption algorithms for GSM and EDGE, and the GEA3 encryption algorithm for GPRS; Document 2: implementors' test data				✓
TS	55.218	Specification of the A5/3 encryption algorithms for GSM and EDGE, and the GEA3 encryption algorithm for GPRS; Document 3: design and conformance test data				✓
TR	55.919	Specification of the A5/3 encryption algorithms for GSM and EDGE, and the GEA3 encryption algorithm for GPRS; Document 4: design and evaluation report				✓

ME = Mobile Equipment; FIGS = Fraud Information Gathering System; IST = Immediate Service Termination; 3G = Third Generation; 3GPP = Third Generation Partnership Project; NDS = Network Domain Security; MAP = Mobile Application Part; IP = Internet Protocol; GPRS = General Packet Radio Service; AF = Authentication Framework; EDGE = Enhanced Data rates for GSM Evolution; GEA3 = GPRS Encryption Algorithm 3

simultaneously, leading to *macrodiversity* that guarantees better quality for the connection. The role of the specific RNC that maintains the connection to the CN (for a particular UE) is called the Serving RNC (SRNC), while another RNC that connects to UE is called the Drifting RNC (DRNC).

In the CN, the most important element on the CS domain is the switching element MSC that is typically integrated with a VLR, which contains a database of the users currently in the location area controlled by the MSC. There is also a Gateway MSC (GMSC) for the purpose of connecting the mobile network to the Public Switched Telephone Network (PSTN). On the packet-switched side the role of

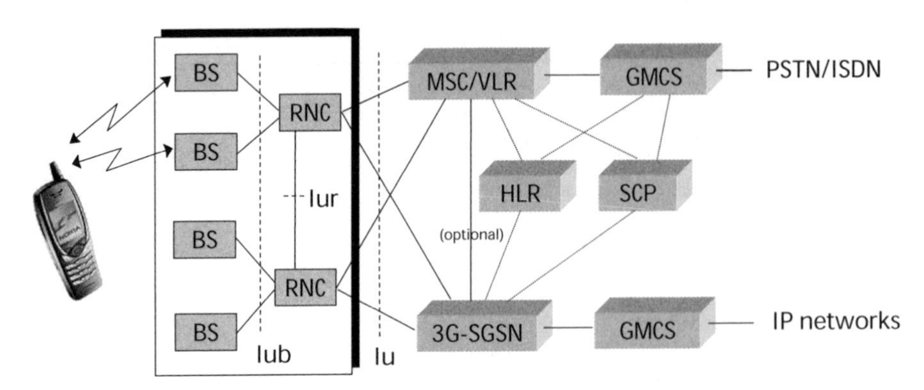

Figure 1.5 3G system architecture based on GSM/GRPS architecture
3G = Third Generation; GSM = Global System for Mobile; GPRS = General Packet Radio Service; MS = Mobile Station (WCDMA); BS = Base Station (WCDMA); RNC = Radio Network Controller; 3G-SGSN = 3G Serving GPRS Support Node; GGSN = Gateway GPRS Support Node; SCP = Service Control Point; HLR = Home Location Register; MSC/VLR = Mobile Switching Centre/Visitor Location Register; GMSC = Gateway MSC; UTRAN = UMTS Terrestrial Radio Access Network; PSTN = Public Switched Telephone Network; ISDN = Integrated Services Digital Network

MSC/VLR is taken by the SGSN, while GGSN takes care of connecting to the outside world (e.g., to the Internet).

In the home network, the static subscriber information is maintained in the Home Location Register (HLR), which is typically integrated with the AuC that holds the permanent security data related to subscribers and creates data that can be used for security features in the serving network and, especially, in the access network.

In addition to the elements listed in this section and illustrated in Figure 1.5, there are many other components in the 3G architecture. One example is the Short Message Service Centre (SMSC), which supports storing and forwarding of short messages. Also, different application-level servers can be added to the system (e.g., Wireless Application Protocol (WAP)), and they can be accessed and utilized over the GPRS-level network.

1.4.2 Protocols in the 3GPP system

The main functionalities in the 3GPP system are:

- Communication Management (CM) for user connections (e.g., call-handling for the CS domain and session management for the PS domain);
- Mobility Management (MM) covers procedures related to user mobility and also security;
- Radio Resource Management (RRM) covers power control for radio connections, control of handovers and system load.

The CM functions are located in the non-access stratum while the RRM functions are located in the access stratum. Unlike the GSM system, MM functions are divided between CN and RAN and the RNC has an active role in MM.

The division of protocols into *user plane* and *control plane* marks an important partition. User plane protocols deal, as the name indicates, with the transition of user data and other directly user-related information (e.g., speech). Control plane protocols are needed to ensure the correct system functionality by transferring necessary control information between the elements in a system. In a telecommunication system there is a further division, *management plane*, which is needed to keep all the elements of a system in operation. Usually, management plane is much less standardized than user plane or control plane.

The dominant protocol used on the Internet is the Internet Protocol (IP), which plays a central role in the PS domain of the 3GPP system. User Datagram Protocol (UDP) and Transmission Control Protocol (TCP) are important as well. In the wireless environment there is a good reason to favour UDP over TCP: because of phenomena like *fading* and temporary loss of coverage it is difficult to maintain reliable transmission of packets on a continuous basis. There is also a 3GPP-specific protocol that is not only run on top of UDP/IP in the PS domain interfaces in CN interfaces but also in the Iu interface. This is the GPRS Tunneling Protocol (GTP). It has been optimized for data transfer along the backbone of the PS domain network.

A central protocol in Session Management (SM) of the PS domain is the Packet Data Protocol (PDP). When SM is active, a PDP context exists, which contains such information as the addresses of the communicating end points and the Quality of Service (QoS) class that is used. There are four different QoS classes defined in the first releases of 3GPP:

- conversational class;
- streaming class;
- interactive class;
- background class.

An illustrative example of the protocol stacks in the 3GPP system is given in Figure 1.6. MM for the CS part and the PS part are handled independently of each other (see Figure 1.7). This has an effect on many security features, as we shall see in subsequent sections of this book. We discuss radio network protocols more closely in the next chapter, mainly because they are closely related security features (the subject matter of Chapter 2).

Let us conclude this section by taking an example that shows how different protocols are typically used. A user receives a phone call. First, the network has to *page* for the user (paging is an MM procedure: the network has to find which geographical area the user lives in). After the user has received the paging message,

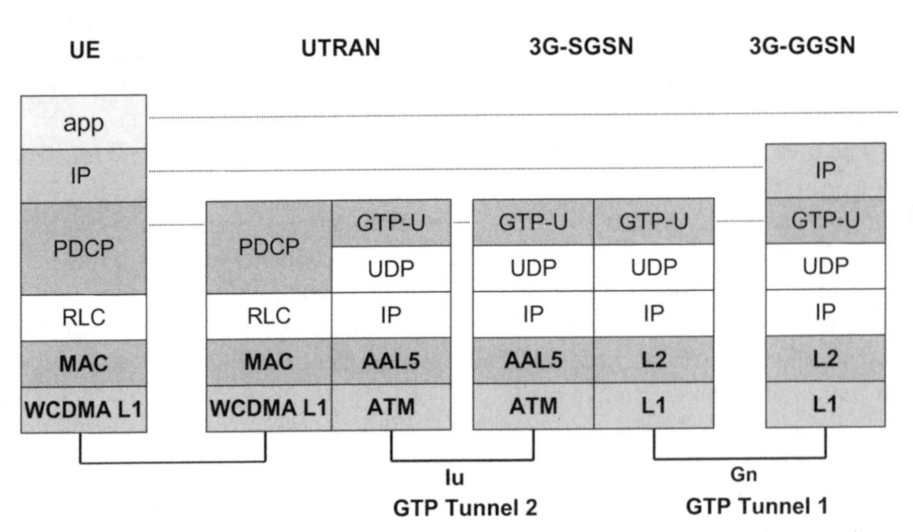

Figure 1.6 Example of a 3GPP protocol stack, showing the 3G packet domain user plane
3G = Third Generation; UE = User Equipment; UTRAN = UMTS Terrestrial Radio Access Network; SGSN = Serving GPRS Support Node; GGSN = Gateway GPRS Support Node; IP = Internet Protocol; PDCP = Packet Data Convergence Protocol; RLC = Radio Link Control; MAC = Medium Access Control; WCDMA = Wideband Code Division Multiple Access; GTP = GPRS Tunneling Protocol; UDP = User Datagram Protocol; AAL = ATM Adaptation Layer; ATM = Asynchronous Transfer Mode

radio connection is established by RRM procedures. When the radio connection is live, there may be an authentication procedure, which once again belongs to MM. Next, the actual call set-up (CM procedure) follows during which, for instance, the user may be informed about who is calling. During the call there may be many further signalling procedures (e.g., for handovers). At the end of the call, the call is first released by a CM procedure and then the radio connection is released by RRM. In between the call, signalling takes place between the CN and the radio network in the same way as between different CNs, depending on where the user is and where (and from whom) the call comes.

1.5 WCDMA Radio Technology

We now take a brief look at the basic concepts of a cellular network and discuss the characteristics of the WCDMA system. For a thorough presentation of WCDMA technology see [56].

The basic idea behind a cellular system is simple: each base station communicates over the radio link with terminals in a restricted area, called the *cell*. When all cells are combined we get the total *area of coverage*. Typically, this area does not give complete geographic coverage, but it covers enough areas to serve most customers in most circumstances.

Networks are carefully planned, which includes finding optimal sizes of cells and

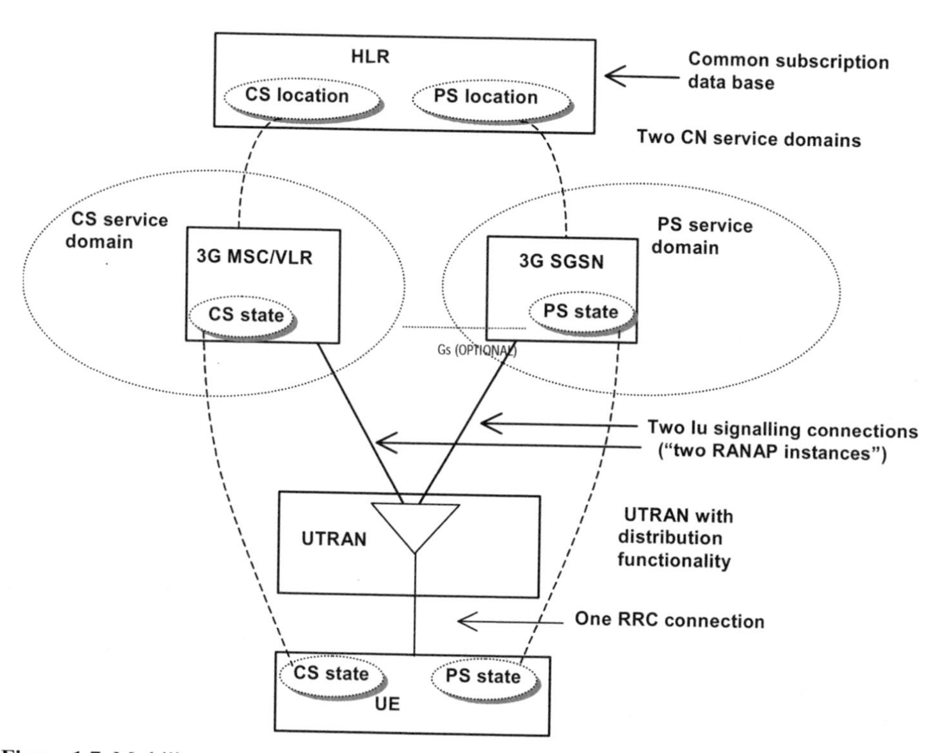

Figure 1.7 Mobility management in 3GPP system: general view on GMM functionality
MM = Mobility Management; HLR = Home Location Register; CS = Circuit Switched; PS = Packet Switched; CN = Core Network; MSC/ULR = Mobile Switching Centre/Visitor Locaton Register; SGSN = Serving GPRS Support Node; RANAP = Radio Access Network Application Protocol; UTRAN = UMTS Terrestrial Radio Access Network; RRC = Radio Resource Control; UE = User Equipment

optimal positions for base stations. Each cell has a limited capacity because it uses certain frequency bands to serve all users inside the cell. Other cells can reuse the same frequencies, implying that by adding new cells to the network more and more subscribers can be served by the system overall. However, there are also limiting factors (e.g., interference both inside a cell and between different cells).

One base station can send and receive signals from several terminals simultaneously. This is guaranteed because of different multiple access techniques. As explained earlier, first generation mobile networks used the FDMA technique in which each user has a dedicated narrow frequency band, allowing the BS to communicate with each user in the cell over different frequencies.

In GSM, the TDMA technique is used in which several users share each frequency and each user controls a short *time slot*, but stays inactive for the rest of the time. As the interval between two consecutive time slots devoted to the same user is short, the user cannot observe the interruptions.

Another dimension further to time and frequency is utilized in the CDMA (Code Division Multiple Access) technique: many users can transmit and receive signals at the same time and frequency, but their signals can still be distinguished because everyone uses different *codes*. These codes are bit strings that have the characteristic that each code and its negative are as different as possible from other codes and their negatives.

1.5.1 CDMA: an example

Let us take a simplified example of the idea underlying CDMA. Say we have three simultaneous users, whose codes are described in the formula (1.1):

$$\begin{array}{lrrrr} C_1 = & 1 & 1 & 1 & 1 \\ C_2 = & 1 & 1 & -1 & -1 \\ C_3 = & 1 & -1 & 1 & -1 \end{array} \tag{1.1}$$

When user i wants to transmit a digit 1, he or she sends instead the whole code string C_i. If the user wants to transmit a digit -1, then he or she sends the negative of C_i. Assume now that these three users want to send the following messages consisting of three digits each:

$$\begin{array}{lrrr} M_1 = & 1 & -1 & 1 \\ M_2 = & 1 & 1 & 1 \\ M_3 = & -1 & -1 & -1 \end{array} \tag{1.2}$$

Now, user 1 actually sends 12 digits as follows: 1 1 1 1 -1 -1 -1 -1 1 1 1 1, and similarly the other two users send their 12 digits. Because they all transmit at the same time and at the same frequencies, the total signal can be found by combining all the individual signals as shown in Formula (1.3):

$$\begin{array}{lrrrrrrrrrrrr} \text{User 1:} & 1 & 1 & 1 & 1 & -1 & -1 & -1 & -1 & 1 & 1 & 1 & 1 \\ \text{User 2:} & 1 & 1 & -1 & -1 & 1 & 1 & -1 & -1 & 1 & 1 & -1 & -1 \\ \text{User 3:} & -1 & 1 & -1 & 1 & -1 & 1 & -1 & 1 & -1 & 1 & -1 & 1 \\ \text{Total signal:} & 1 & 3 & -1 & 1 & -1 & 1 & -3 & -1 & 1 & 3 & -1 & 1 \end{array} \tag{1.3}$$

In this simplified example, signals are combined by simply adding the signal values together. This method really needs more justification but it serves the illustrative purpose here.

At the receiving end in our example of the network side the received signal is compared with the code strings. This is done in "slices" of 4 digits each. We try to find out whether there is a better correlation of the received signal with the code or with the negative of the code. Mathematically, we can compute the *inner product* of each code and the received signal. For code C_1 the inner products are as follows:

$$
\begin{aligned}
&1 \cdot 1 + 1 \cdot 3 + 1 \cdot (-1) + 1 \cdot 1 = 4, \\
&1 \cdot (-1) + 1 \cdot 1 + 1 \cdot (-3) + 1 \cdot (-1) = -4 \\
\text{and} \quad &1 \cdot 1 + 1 \cdot 3 + 1 \cdot (-1) + 1 \cdot 1 = 4
\end{aligned} \tag{1.4}
$$

Now, better correlation with the code is indicated by a positive result, while better correlation with the negative of the code is seen as a negative result of computation. Thus, the receiving side may decode the sent message string M_1. Respectively, for the code C_2 we get the inner products as follows:

$$
\begin{aligned}
&1 \cdot 1 + 1 \cdot 3 + (-1) \cdot (-1) + (-1) \cdot 1 = 4, \\
&1 \cdot (-1) + 1 \cdot 1 + (-1) \cdot (-3) + (-1) \cdot (-1) = 4 \\
\text{and} \quad &1 \cdot 1 + 1 \cdot 3 + (-1) \cdot (-1) + (-1) \cdot 1 = 4
\end{aligned} \tag{1.5}
$$

The transmitted digits in M_2 are all 1's because the results are all positive. Finally, for the third user we compute:

$$
\begin{aligned}
&1 \cdot 1 + (-1) \cdot 3 + 1 \cdot (-1) + (-1) \cdot 1 = -4, \\
&1 \cdot (-1) + (-1) \cdot 1 + 1 \cdot (-3) + (-1) \cdot (-1) = -4 \\
\text{and} \quad &1 \cdot 1 + (-1) \cdot 3 + 1 \cdot (-1) + (-1) \cdot 1 = -4
\end{aligned} \tag{1.6}
$$

Hence, M_3 exclusively consists of -1's.

This simplified example illustrates how it is possible to sort out independently-sent signals even if they are transmitted on top of each other. In practice, there are many complications that are not shown in this example. Most notably, there is no easy way to guarantee that all users transmit their codes in a fully-synchronized way. This is why the codes for uplink traffic have to be designed in such way that correlations can be observed, even in the case where transmission is not fully synchronized. On the other hand, for *downlink* traffic full synchronization is possible, since the sending entity is in the network. This is the main reason why codes used in the uplink direction are different from the codes used in the opposite direction.

1.5.2 Basic facts of WCDMA

In UMTS there are two variants of WCDMA technology: Frequency Division Duplex (FDD) and Time Division Duplex (TDD). The difference is in the way uplink and downlink transmissions are separated from each other. In FDD, uplink transmissions use different frequencies from those used for downlink transmissions. The following frequencies are allocated for these purposes by the ITU: 1,920–1,980 MHz for uplink and 2,110–2,170 MHz for downlink. Hence, the *duplex distance* is 190 MHz. In TDD, both uplink and downlink transmissions use the same frequencies but different time intervals. There are two frequency bands allocated for the TDD variant: 1,900–1,920 MHz and 1,980–1,995 MHz.

The first implemented UMTS networks utilize the FDD variant of WCDMA, and we will restrict ourselves to this case for the rest of the book. However, the same security features apply to both FDD and TDD. Note for comparison that the GSM 1800 system uses frequencies of 1,710–1,785 MHz for uplink and 1,805–1,880 MHz for downlink (a duplex distance of 95 MHz). It is also important to note that in the USA frequencies around 1,900 MHz are used for this system and hence the frequencies mentioned above cannot be applied for WCDMA in the USA.

The frequency band of WCDMA is further divided into several slices that are used for different channels. The width of each slice is 5 MHz, containing 3.84 MHz of *effective* bandwidth and *guard bands* around it.

User data is given as a string of bits (0 and 1) that are coded into digits (1 and −1). As illustrated in the example above, one digit in user data corresponds to several digits in the code that is used for *spreading*. To make a clear distinction between these two types of digits, the latter is called a *chip* instead. In WCDMA the *System Chip Rate* is 3.84 Mchip per second.

There are several types of codes that are used simultaneously: during uplink, the *scrambling code* separates users from each other, the *channelization code* separates different channels allocated to the same terminal and the *spreading code* is the product of the two former codes; during downlink, the scrambling code separates cells from each other, the channelization code separates users in the same cell from each other and the spreading code is again the product of these two codes. These codes are mathematically fairly complex and it is not obvious how an outside observer would discover which codes are in use at a given moment in time. It may be thought that this kind of complexity increases the security level of the system because it is difficult for an intruder even to find out what is going on in the WCDMA radio frequencies. However, this merely implies that the intruder has to utilize fairly sophisticated equipment, but various kinds of test equipment are widely available.

WCDMA divides periods of time into *frames*. One WCDMA frame lasts 10 ms and is further divided into 15 time slots. There is a special bit structure in each time slot, beginning with *pilot* bits followed by bits that support power control. Of course, actual data bits constitute the main portion of the time slot. To correct transmission errors, *channel coding* is used in WCDMA (as is already done in GSM). In WCDMA, *convolutional* codes and *Turbo* codes are used to correct errors. They are specified (for FDD) in the 3GPP specification 25.212 [6].

The WCDMA utilizes the fact that the transmitted signal propagates through several different paths, effected by means of special types of receivers, such as the

Rake receiver that contains several *fingers*, each of which receives part of the total signal.

1.5.3 Handovers

The concept of a handover is crucial for a cellular system. A basic handover process consists of three phases:

- *measurement phase*, where both the UE and the BS continuously do measurements (e.g., signal quality);
- *decision phase*, where certain criteria are used in order to decide that a handover is to be carried out;
- *execution phase*, where the UE actually changes the cell.

In addition to the co-operation between terminal and the network in taking measurements, the decision about carrying out a handover can also be taken by the UE and by the network. However, even in a *mobile-initiated* handover the final decision is done by the RNC, because it is responsible for the overall management of radio resources in the system. For *network-initiated* handovers, the decision is typically taken by serving RNC, but for traffic reasons the decision can also be taken by the MSC.

There are two basic handover types: *hard* and *soft*. In the former, the connection in the old cell is released before the new connection in the new cell is established. This is the type of handover that is also dominant in 2G systems. It is also possible to execute hard handovers between the GSM and the WCDMA (Figure 1.8).

In soft handovers the old connection is released *after* the new connection has already been established. Therefore, for a period of time macrodiversity is utilized. The mechanisms to combine the signals are quite different in this case, compared with those in the microdiversity case, because the combination takes place in the RNC instead of the BS. The majority of WCDMA handovers are soft handovers (Figure 1.9).

1.5.4 Power control

In the WCDMA system, several users transmit (and receive) at the same frequency. From each user's point of view, the signals of other users are basically treated as interference. Because of this, it is vital that the power levels used by different users are well balanced. From the BS point of view, the power level of a received uplink

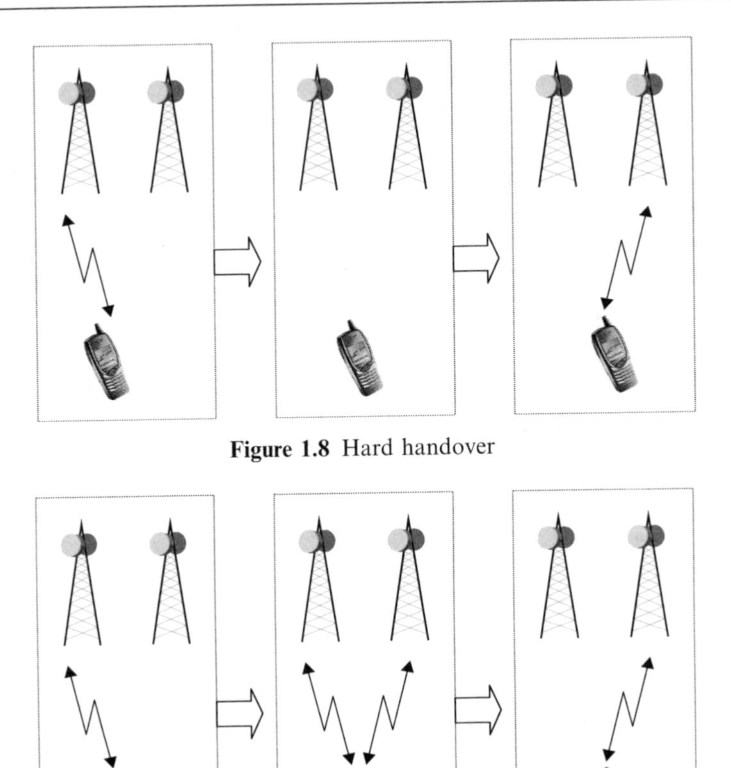

Figure 1.8 Hard handover

Figure 1.9 Soft handover

signal from each UE should be the same regardless of the distance between UE and the BS. UEs are constantly on the move and, therefore, the power levels have to be adjusted continuously.

Power control is not as essential for downlink as it is for uplink. The reason for this is simple: only one relevant element is moving—the receiving UE. Still, power control is also utilized in downlink to minimize interference from other cells. In any event, the mechanisms to ensure downlink power control are simpler than the mechanisms for uplink.

The importance of power control in WCDMA system can be highlighted by comparison with the corresponding feature in GSM. In the GSM system, the power level of the connection is adjusted once or twice in a second, whereas in WCDMA the power control is applied 1,500 times per second.

There are two basic types of power control methods. In Open Loop Power Control (OLPC) the UE adjusts its transmission power based on the received (pilot) signal level from the BS. This happens typically when the UE is in idle

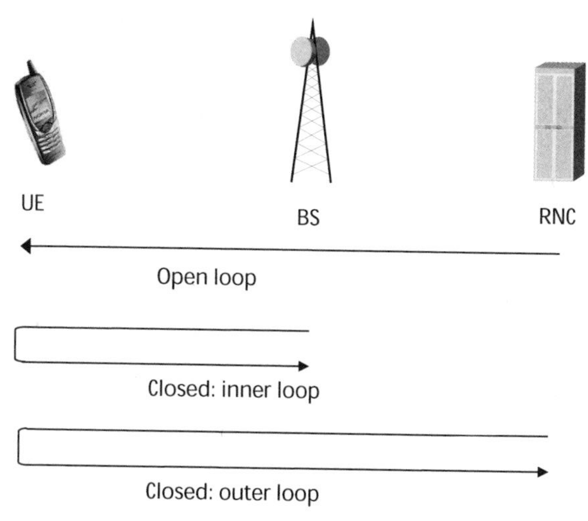

Figure 1.10 Power control mechanisms
RNC = Radio Network Controller

mode. There may be drastic and rapid fading in the radio channel and so OLPC is insufficient on its own as a power control mechanism (Figure 1.10).

The Closed Loop Power Control (CLPC) is utilized when radio connection has been established. The method works in this way: the network side measures the received signal from the UE and feeds back information about whether the UE needs to increase or decrease its transmission power level. If the feedback comes from the BS, then we talk about Inner Loop Power Control (ILPC) or Fast Power Control (FPC). In outer CLPC the decision is taken by the RNC. In addition to signal strength, the BS takes other measurements (e.g., Signal-to-Interface Ratio (SIR) and Bit Error Rate (BER)).

At first sight it may look as if the basic characteristics of CDMA make it more robust against intruders: a degree of tolerance for radio interference is built in because even legal users interfere with each other. On the other hand, the complexity of the power control means that in case the power control somehow fails there are serious consequences for system performance. In any event, the jamming of radio frequencies seems to be an ongoing threat in every cellular system.

2

UMTS Security Features in Release 1999

2.1 Access Security to UMTS

Radio access technology will change from TDMA (Time Division Multiple Access) to WCDMA (Wideband Code Division Multiple Access) when the Third Generation (3G) mobile networks are introduced. Despite this shift, requirements for access *security* will not change. It is an absolute prerequisite of UMTS (Universal Mobile Telecommunications System) that end-users of the system are *authenticated* (i.e., the identity of each subscriber is verified): nobody wants to pay for fraudulent calls that are made by others.

The *confidentiality* of voice calls is protected in the Radio Access Network (RAN), as is the confidentiality of transmitted user data. This means that the user has control over choosing the parties he or she wants to communicate with. Users also want to *know* that confidentiality protection is really applied and so *visibility* of applied security mechanisms is needed. *Privacy* of a user's whereabouts is generally appreciated; most of the time an average citizen does not care whether it is possible to trace where he or she is, but if persistent tracking occurs the user would rightly be irritated. Similarly, precise information about the location of people would be useful to burglars. The privacy of user data is another issue that is critical during transfer through the network (privacy and confidentiality are largely synonymous in this presentation).

UMTS *accessibility* is clearly important for subscribers who are paying for it, but network operators consider *reliability* of network functionality to be equally important: they need control within network functions to be effective. This is guaranteed by the *integrity* of radio network signalling, which checks that all control messages have been created by authorized elements of the network. In general, integrity checking protects against any manipulation of a message (e.g., insertion, deletion or substitution).

The most important ingredient in providing security for network operators and subscribers is *cryptography*, which consists of various techniques that have their roots in the science and art of *secret writing*. It is sometimes useful to make communication deliberately incomprehensible (i.e., using ciphers or, synonymously, encryption). This is the most effective way to protect communications against eavesdroppers. Cryptographic issues are thoroughly discussed in Part II.

In the present chapter, we go through the security features introduced in the first release of the 3GPP system specifications (Release 1999).

2.1.1 Mutual authentication

There are three entities involved in the authentication mechanism of the UMTS system:

- Home Environment (HE);
- Serving Network (SN);
- terminal, more specifically USIM (Universal Subscriber Identity Module), typically in a smart card.

The basic idea is that the SN checks the subscriber's identity (as in GSM—Global System for Mobile communications) by a *challenge-and-response* technique while the terminal checks that the SN has been authorized by the home network to do so. The latter part is unique to UMTS (not available with GSM) and through it the terminal can check that it is connected to a legitimate network.

The mutual authentication protocol itself does not prevent the active attack scenario of Figure 1.1, but in combination with other security mechanisms it guarantees that the active attacker cannot get any real benefit out of the situation. The only possible gain for the attacker is to be able to disturb the connection (but an attacker could also do this by means of radio-jamming). At the moment no protocol method can circumvent such an attack.

The cornerstone of the authentication mechanism is a *master key* or a subscriber authentication key K, which is shared between the USIM of the user and the home network database, Authentication Centre (AuC). The key is permanently kept secret and has a length of 128 bits. The key K is never transferred from these two locations (i.e., the user has no knowledge of the master key).

Apart from mutual authentication, keys for encryption and integrity checking are also derived. These are temporary keys (with the same length of 128 bits) and are derived from the permanent key K during every authentication event. It is a basic principle in cryptography to keep the use of permanent keys to a minimum and, instead, derive temporary keys from it for protection of bulk data.

We now describe the Authentication and Key Agreement (AKA) mechanism at a general level. The design of the mechanism was begun by combining two different

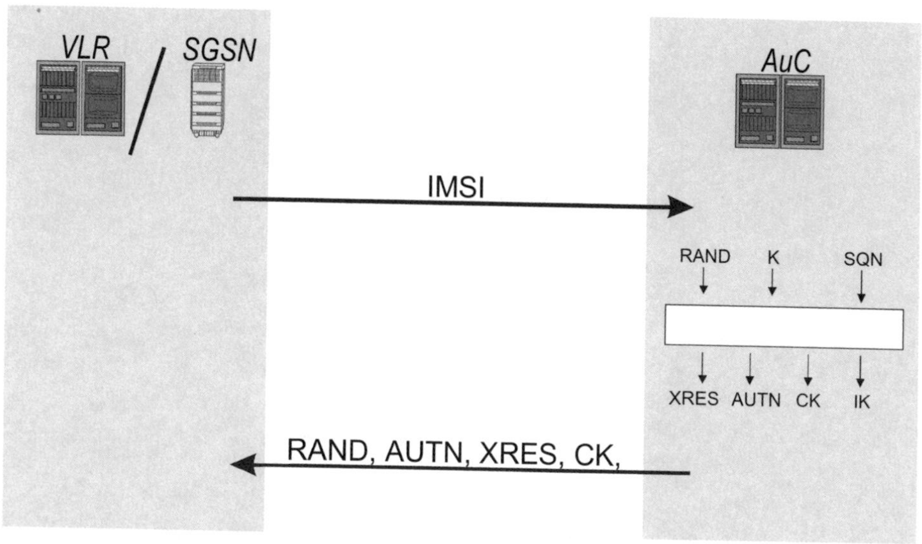

Figure 2.1 Authentication data request and authentication data response
VLR = Visitor Location Register; SGSN = Serving GPRS Support Node; AuC = Authentication Centre; IMSI = International Mobile Subscriber Identity; RAND = random number; SQN = sequence number; XRES = expected response; AUTN = authentication token; CK = Cipher Key; IK = Integrity Key; GPRS = General Packet Radio Service

authentication mechanisms: GSM's authentication and key agreement mechanism [29] and a generic authentication mechanism based on sequence numbers specified in an ISO standard [63].

The authentication procedure begins when the user is identified in the SN. Identification occurs when the identity of the user (i.e., permanent identity International Mobile Subscriber Identity (IMSI), or temporary identity Temporary Mobile Subscriber Identity (TMSI), or Packet TMSI (P-TMSI)), has been transmitted to the VLR (Visitor Location Register) or SGSN (Serving GPRS Support Node). Then the VLR or SGSN sends an *authentication data request* to the AuC in the home network.

The AuC contains the master key of each user and, based on the knowledge of IMSI, the AuC is able to generate *authentication vectors* for the user. The generation process contains executions of several cryptographic algorithms, which are described in more detail in Chapter 8. The generated vectors are sent back to the VLR/SGSN in the *authentication data response*. This process is depicted in Figure 2.1. These control messages are carried on the MAP (Mobile Application Part) protocol.

In the SN, one authentication vector is needed for each authentication instance (i.e., for each run of the authentication procedure). This means that the (potentially-long distance) signalling between SN and AuC is not needed for every authentication event and that in principle this signalling can be done independently of user actions after initial registration. Indeed, the VLR/SGSN may fetch new authentication vectors from AuC well before the number of stored vectors runs out.

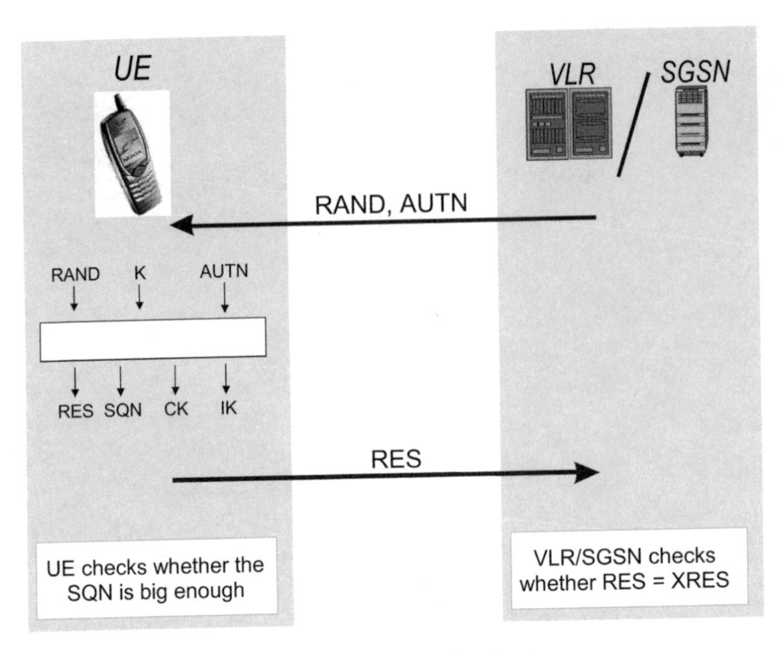

Figure 2.2 User authentication request and user authentication response
UE = User Equipment; VLR = Visitor Location Register; SGSN = Serving GPRS Support Node; RAND = random number; AUTN = authentication token; RES = user response; SQN = sequence number; CK = Cipher Key; IK = Integrity Key; XRES = expected response; GPRS = General Packet Radio Service

The SN (VLR or SGSN) sends a *user authentication request* to the terminal, containing two parameters from the authentication vector, called RAND and AUTN. These parameters are transferred to the USIM, which exists inside a tamper-resistant environment (i.e., in the Universal Integrated Circuit Card—UICC). The USIM contains the master key K and, using it with the RAND (random number) and AUTN (authentication token) parameters along with other input values, USIM carries out a computation that resembles the generation of authentication vectors in AuC. This process also involves running several algorithms, just as in the corresponding AuC computation. The result of the computation gives the USIM the ability to verify whether the AUTN parameter:

- was indeed generated in AuC;
- was not sent beforehand to the USIM.

In the positive case, the computed RES parameter is sent back to the VLR/SGSN as part of the *user authentication response*. Now, the VLR/SGSN is able to compare the user response (RES) with the expected response (XRES), which is part of the authentication vector. If they match, authentication ends positively. This part of the process is depicted in Figure 2.2.

The keys for Radio Access Network (RAN) encryption and integrity protection (namely, Cipher Key (CK) and Integrity Key (IK)) are created as a by-product in the authentication process. These temporary keys are included in the authentication vector and, thus, are transferred to the VLR/SGSN. These keys are later transferred to the Radio Network Controller (RNC) in the RAN when encryption and integrity protection start. Respectively, the USIM is able to compute the CK and IK after it has obtained the RAND (and verified it through the AUTN). Temporary keys are subsequently transferred from USIM to the Mobile Equipment (ME) where the encryption and integrity protection algorithms are implemented.

In the following sections we take a more detailed look at the mechanisms needed for authentication and key agreement.

2.1.1.1 Authentication vector generation

We now take a closer look at the generation of authentication vectors in the AuC. An illustration of the process is given in Figure 2.3. The process begins by picking an appropriate sequence number (SQN). Roughly speaking, what is required is that SQNs are chosen in ascending order. A more detailed description about how to create SQNs is given in Section 2.1.1.3. The purpose of the SQN is to provide the user (or more technically the USIM) with proof that the generated authentication vector is *fresh* (i.e., it has not been used before in an earlier run of authentication). In parallel with the choice of SQN, a 128-bit long RAND is generated. This is a

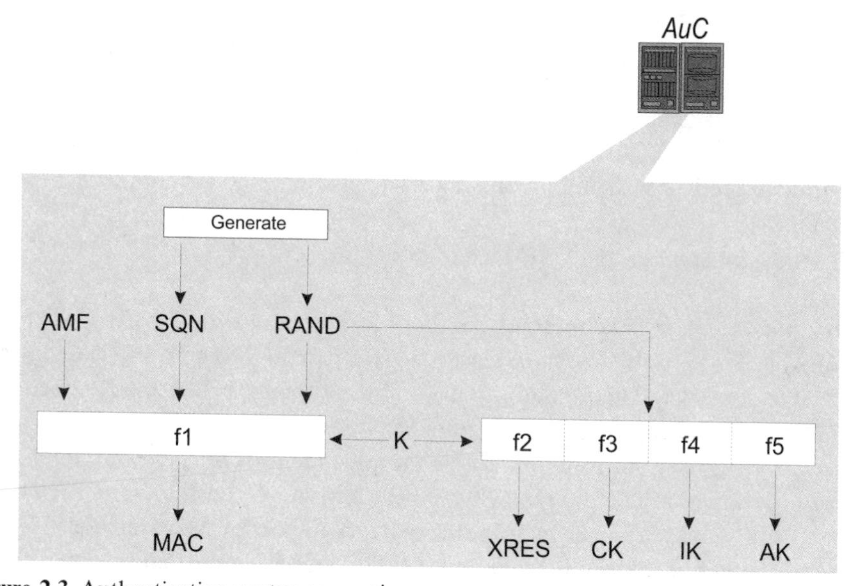

Figure 2.3 Authentication vector generation
AuC = Authentication Centre; AMF = Authentication Management Field; SQN = sequence number; RAND = random number; MAC = Message Authentication Code; XRES = expected response; CK = Cipher Key; IK = Integrity Key; AK = Anonymity Key

demanding task in itself, but in this presentation we just assume that a cryptographic pseudorandom generator is in use that is able to produce large amounts of unpredictable output bits, when a good physical random source is available to produce smaller amounts of random bits that can be used as an input (seed) for the pseudorandom generator.

The key concept in authentication vector computation is a mathematical function, called *one-way function*, which is relatively easy to compute but practically impossible to invert. In other words, as long as we have input parameters there exists a fast algorithm to compute output parameters, but if the output parameters are not known, then there exist no efficient algorithms to deduce any input that would produce the output. Of course, there is a simple algorithm, called the exhaustive search algorithm, that can be used to find the correct input by trying all possible choices until one gives the requisite output. However, this algorithm quickly becomes extremely inefficient as the length of input increases.

In total, five one-way functions are used to compute the authentication vector. These functions are denoted by f1, f2, f3, f4 and f5. The function f1 differs from the other four in that it takes four input parameters: master key K, RAND, SQN and finally an administrative Authentication Management Field (AMF). All other functions from f2 to f5 only take K and RAND as inputs. The requirement of the one-way property is common to all functions f1–f5. They can all be built around the same *core* function. However, it is essential that they differ from each other in a fundamental way so that the output of one function reveals no information about the outputs of the other functions. The output of f1 is Message Authentication Code (MAC) (64 bits) and the outputs of f2, f3, f4 and f5 are, respectively, XRES (32–128 bits), CK (128 bits), IK (128 bits) and AK (64 bits). The authentication vector consists of the parameters RAND, XRES, CK, IK and the authentication token (AUTN). The last one is obtained by concatenating three different parameters: SQN added bit by bit to AK, AMF and MAC. All of the functions involved in the AKA procedure are studied in detail in Chapter 8 of this book.

2.1.1.2 Authentication on the USIM side

We now take a closer look into the handling of authentication on the USIM side (illustrated in Figure 2.4). The same functions f1–f5 are involved on this side but in a slighty different order. The function f5 has to be computed before the f1, since f5 is used to conceal the SQN. This concealment is needed in order to prevent eavesdroppers from getting information about the user identity through the SQN. The output of the function f1 is marked XMAC (or XMAC-A) on the user side. This is compared with the MAC received from the network as part of the parameter AUTN. If there is a match it implies RAND and AUTN have been created by some entity that knows K (i.e., the AuC of the user's home network).

Of course, there is still the possibility that some attacker who has recorded an

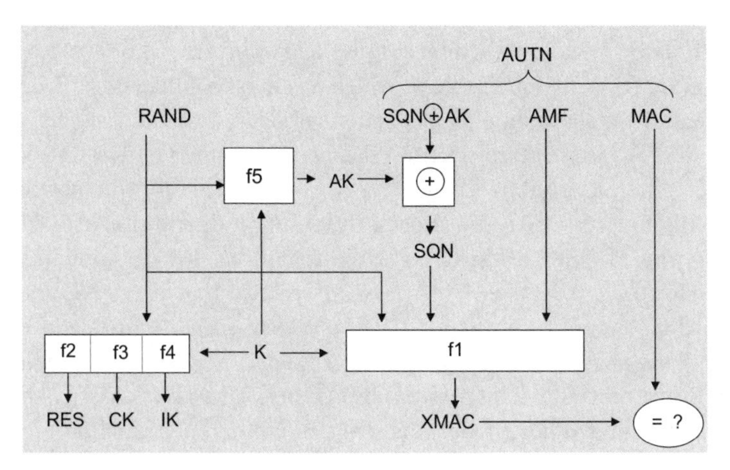

Figure 2.4 Authentication handling in USIM
AUTN = authentication token; RAND = random number; SQN = sequence number; AK = Anonymity Key; AMF = Authentication Management Field; MAC = Message Authenication Code; RES = user response; CK = Cipher Key; IK = Integrity Key; XMAC = expected MAC

earlier authentication event could ascertain the RAND and AUTN. However, as mentioned above, the SQN protects against this threat. The USIM should simply check that it has not seen the same SQN before, and the easiest way to do this is to require that SQNs appear in ascending order. It is also possible for the USIM to allow some SQNs to arrive out of order (e.g., by maintaining a shortlist of the greatest SQNs received so far). In the next sections we will take a closer look at this issue.

Since the transfer of authentication vectors from the AuC and the actual use of these vectors for authentication are done somewhat independently, there are several reasons why it is possible that authentication vectors may be used in a different order from which they were originally generated. The most obvious reason for this is because of the fact that mobility management functions for the CS (Circuit Switched) and PS (Packet Switched) domain are independent of each other, implying that authentication vectors are fetched to the VLR and SGSN independently of each other and that the vectors are also used independently.

The choice of algorithm (f1–f5) is in principle operator-specific, because they are only used in the AuC and in the USIM and the same home operator controls both of these entities. An example set of algorithms (called MILENAGE) exists in the Third Generation Partnership Project (3GPP) specification TS 35.206 [24] (these algorithms are discussed thoroughly in Chapter 8).

2.1.1.3 SQN generation in the AuC

SQN management is also operator-specific in principle. There are two basic strategies at work in creating SQNs: each user may have an individual SQN, or SQN generation may be based on a global counter (e.g., universal time). A combination of these two strategies is also possible in which the most significant part of the SQN is user-specific but the least significant part is based on a global counter.

In the 3GPP specification 33.102 [9] there is an informative annex C that describes three different options for generating SQNs. Because this part of the specification is only for informative purposes, the network operator is also free to choose some other way of generating SQNs while remaining fully compliant with 3GPP standards. However, it has been observed in practice that excessive diversity inside one standard tends to lead in the long run to interoperability problems of some sort or another. This observation is by no means limited to security mechanisms.

Let us discuss this important issue a bit further. There is a widely-held agreement inside the 3GPP that different optional functionalities for the same purpose in the same standard should be avoided if ever possible. In standardization specification work a decision often has to be made between two (or more) proposed solutions that have equal technical merits but are simply different ways of achieving the same goal. An easy way out from such a situation is to allow different options in the standard. At first sight it may look like the only penalty that has to be paid for such a compromise decision is the risk that some elements in the system may have to contain redundant, duplicate functionality. However, when viewed in depth there is the much bigger issue of future specifications. It may happen that a new functionality is designed on top of the old functionality for which several implementation options were allowed. As a consequence it becomes difficult to stop these options from being available to the new functionality, which may also be dependent on a number of other old functionalities that again may well contain several options. So, it is difficult to keep the design of the new functionality simple in such cases.

This general concern certainly applies to our context because the UICC manufacturers and AuC manufacturers are usually (if not always) different companies. Also, the issue with future standards has emerged, as the AKA mechanism has been introduced into new contexts in 3GPP Releases 5 and 6 (see Chapter 3). For these reasons, it can be anticipated that the example mechanisms for SQN management presented in [9] are likely to be adopted widely in practice.

Let us now give a brief description of the example mechanisms. However, for full details see annex C in [9]. The SQN for a certain user contains two concatenated parts: SQN = SEQ || IND. The least significant part (5 bits) IND is used to allow effective mechanisms in the USIM side to verify the freshness of the SQN parameter. The general rule is that the IND value is incremented by one for each new authentication vector to be generated. This increment is understood cyclically (i.e., when the IND parameter reaches the maximal value then the next value to be chosen is zero).

It is possible, although not usually the case, that the AuC gets information about the type of node requesting the vector (e.g., whether it is a MSC/VLR or a SGSN). When this happens, it may be useful to differentiate the range of IND values allocated to nodes in each different domain. For instance, IND values that are

Table 2.1 Partitioning the IND value space (an example)

Access to domain (AV sent to)	IND value range
CS domain (MSC/VLR)	0–9
PS domain (SGSN)	10–19
IMS domain (S-CSCF)	20–24
WLAN domain	25–28
Other domains	29–31

AV = Authentication Vector; IND =least significant part of SQN; CS = Circuit Switched; MSC/VLR = Mobile Switching Centre/Visitor Location Register; PS = Packet Switched; SGSN = Serving GPRS Support Node; IMS = IP Multimedia CN Subsystem; IP = Internet Protocol; CN = Core Network; S-CSCF = Serving Call Session Control Function; WLAN = Wireless LAN; GPRS = General Packet Radio Service; LAN = Local Area Network

even are allocated to nodes in the CS domain and odd IND values are allocated to nodes in the PS domain. Consequently, in this example, for two consecutive authentication vectors allocated to the same domain, the difference in IND value would be 2 instead of 1. As mentioned earlier, according to the more recent releases of 3GPP standards, authentication vectors may also be consumed in other domains (e.g., by IMS (IP Multimedia Core Network Subsystem) (Release 5) or by an interworking WLAN (Wireless Local Area Network) system (Release 6)). For effective handling of SQNs in even these cases, it may be useful to introduce a more fine-grained partition of the IND value space. How this partition should exactly be done is highly dependent on the structure of the network in question and the optimal partition probably changes as the network evolves. To elaborate on this a bit more, let us look at the way of partitioning the IND value space as given in Table 2.1.

Authentication Vectors (AVs) may be sent to the destination in *batches*. This reduces the number of times the AuC has to be accessed. At the same time, there is an increased probability that AVs are consumed in a different order than they were generated. Typically, all AVs in a batch share the same value of SEQ and only differ in the value of IND. There are three different strategies used to generate the value of SEQ:

1. SEQ is an individual counter and its current value is maintained in a database independently for each user.

2. SEQ is based on a global counter and for each user a deviation from the global counter, called DIF (difference), is maintained in a database. Ideally the DIF value is 0 for all users, but because of synchronization errors (see Section 2.1.1.5) it may have to be updated for some users.

3. SEQ has two parts: SEQ = SEQ1 | SEQ2, where SEQ1 is an individual counter and SEQ2 is based on a global counter (GLC) that represents universal time. The value of SEQ is maintained in the database for each user in this case as well.

All three ways of generating SEQ are described in [9]. Here we only present the third strategy because the other two can be seen (more or less) as extreme cases of the combined case 3.

A suggested length for parameter SEQ2 is 24 bits, leaving 19 bits for the individual counter SEQ1 as IND consists of 5 bits (the length of a SQN is fixed at 48 bits). The GLC also consists of 24 bits. Using time units of 1 second, the GLC would wrap around once in 194 days. This ensures that almost all users would be authenticated at least once during each GLC period.

The idea is to keep the most significant part (SEQ1) constant until SEQ2 wraps around. The latter correlates heavily with GLC, and therefore the wrap-around would typically happen once in 194 days. We cannot assume that SEQ2 = GLC because it is possible that two batches of authentication vectors are fetched for the same user exactly at the same time (or at least during the same time unit of GLC). Remember that different domains consume and fetch AVs from the AuC independently of each other, and therefore fetching can certainly happen simultaneously. If another fetch occurs during the same time unit as the previous fetch then SEQ2 is incremented by 1 anyway. As a result, SEQ2 would become temporarily greater than GLC in the second fetch.

We can safely assume that the GLC clock rate is on average faster than the authentication frequency for any one specific user. Therefore, even if the SEQ2 temporarily overtakes the GLC, the latter catches up fairly quickly.

Let us assume that the AuC gets a request for a batch of AVs. First, the previous value of SEQ is retrieved from the database (for this particular user) and then the previous value of SEQ2 is compared with the current GLC value. We have three possible cases (see also Figure 2.5):

1. SEQ2 < GLC. This is the usual case, with the new SEQ2 value set to be equal to the current GLC value while the SEQ1 value remains unchanged;

2. GLC < SEQ2, but any difference is small (or even zero). This is the case discussed above (i.e., the previous generation of an AV and, consequently, the previous update of the SEQ2 value have happened very recently or there have been many updates in the very recent past). Here SEQ2 is incremented by 1 and SEQ1 remains unchanged unless there is a wrap-around of SEQ2 as a result of this increment, in which case the SEQ1 is also incremented by 1;

3. GLC < SEQ2, and the difference is large. This is the case where a wrap-around of GLC has occurred since the generation of the last batch of AVs. Here the new

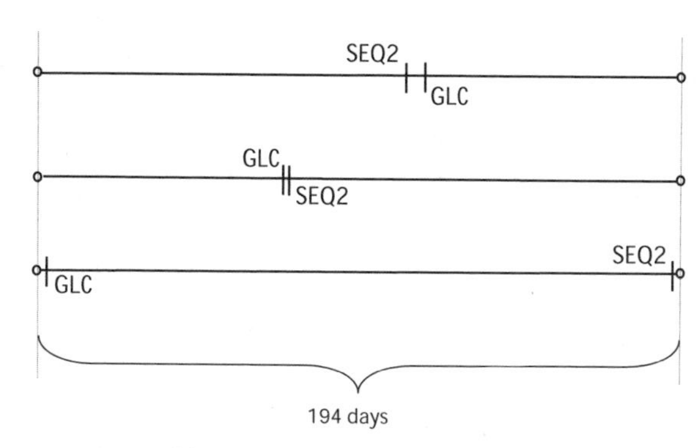

Figure 2.5 SEQ2 update cases
SEQ2 = least significant part of SEQ; GLC = global counter

SEQ2 value is set to be equal to the current GLC value while the SEQ1 value is incremented by 1.

Obviously a precise threshold value is needed to differentiate between cases 2 and 3 and a recommended value is given in [9]: a difference is deemed small if (and only if) it is smaller than 2^{16}. Note that if many, almost simultaneous AVs were accidentally categorized to class 3 instead of class 2, then the only disadvantage is that SEQ1 would be incremented unnecessarily. This does not, however, lead to SQN values being reused. In the opposite case, where the user is inactive for a very long period (say, almost a full period of 194 days) and, as a consequence, AV generation falls into class 2 instead of class 3, there is a small risk that some SQN values could be reused, even though this is never supposed to happen. Anyway, there is no *security* risk involved, since reusing SQN values only implies that network authentication may fail on the USIM side and recovery from this kind of situation is guaranteed by the resynchronization procedure.

2.1.1.4 SQN checking in USIM

As mentioned earlier, SQNs exist in the AKA concept for one reason: they allow the USIM to check whether the authentication challenge has been received before. In addition to the highest SQN received so far, the USIM maintains an array of other received SQNs. The array is indexed by the elements within the range of the IND parameter. In our example case where the length of the IND parameter is 5 bits, consequently there are 32 elements in the array. The element in the array indexed by i is the highest SQN received so far with IND = i.

Let us suppose the USIM receives an SQN = SEQ || IND. This can be compared with the element in the maintained array that is indexed by the value of IND.

If the received SQN is greater than the value in the array, then the SQN is accepted, the network authentication succeeds and the element in the array is replaced by the received value SQN. Clearly, it is enough just to store the SEQ part, because the IND part is indicated by the index of the element in the array.

On the other hand, if the received value is smaller or equal to the value in the array, then the received SQN is not accepted and the network authentication fails. The array element is not changed, but, instead, a resynchronization procedure is initiated.

In addition to the comparison mentioned above, the received SQN value is also checked against the highest SQN value stored in the USIM. The purpose of this check is to prevent arbitrarily-big jumps in SQN values. Therefore, the received SQN value is not accepted unless it increases the highest SQN value by at most a value of Δ. If this check is not done, there is a small chance that the USIM gets into a situation where it has accepted and stored such large values of SQN that the only way out is to do a wrap-around of these counters. The limit value Δ has to be chosen carefully in order to make sure that normal jumps in SQN are not considered abnormal. An example value $\Delta = 2^{28}$ is given in [9].

2.1.1.5 Synchronization of SQNs

The mutual authentication mechanism is based on two parameters that are stored in both the AuC and USIM: a static master key K and a dynamic SQN. It is vital that these parameters are kept synchronized on both sides. For the static K this is easy, but it is possible for dynamic SQNs to get out of synchronization for some reason. As a consequence, authentication would fail. A specific *resynchronization procedure* is used in this case (see Figure 2.6). By using the master key K as the basis for secure communication, the USIM informs the AuC of its current (highest) SQN value.

The AUTS parameter is delivered during resynchronization. It contains two parts: the sequence number of the USIM concealed by AK and a message authentication code MAC-S computed by another one-way function f1* from the input parameters SQN, K, RAND and AMF. The last two parameters are obtained from the failed authentication event. The one-way function f1* has to be different from f1 because, otherwise, already recorded AUTN parameters could in principle be accepted as valid AUTS parameters in the resynchronization and an attacker could at least disturb the authentication process. When the AuC receives the AUTS parameter, it carries out the following steps:

1. The SQN_{USIM} is computed from AUTS.
2. Based on the value of SQN_{USIM}, the AuC checks whether the next authentication vector would be acceptable to the USIM—
   a if YES, then the process continues from step 4;
   b if NO, then

3. The AuC checks whether the the MAC-S value in AUTS is correct—
   a if YES, then the value SQN_{AuC} is reset to SQN_{USIM} and the process continues from step 4;
   b if NO, then SQN_{AuC} is not reset but the process continues anyway from step 4.
4. The AuC sends a batch of fresh AVs to the VLR/SGSN.

Note that new AVs are also sent to the SN node when the MAC-S value is either not checked (2a) or failed the check (3b). However, a general cryptographic principle states that no action should be taken when a message authentication code turns out to be false. Nevertheless, in our case the sending of new AVs is justified for the following reasons:

- if the AUTS parameter was computed by the genuine USIM, the UE would then try to get access to the network again after the first attempt has failed, resulting in a new AV being needed from the AuC in any event;
- if the AUTS parameter was computed and sent to an attacker (for whatever reason), the attacker might try to access the network again, but once more a new AV is probably fetched from the AuC.

Equipped with the new AVs, the VLR/SGSN is able to authenticate the UE in case he or she tries to get access again. If the UE repeatedly indicates network authentication failures by sending more AUTS values, the two most probable reasons are:

- there is something wrong with the computations or data in the USIM;
- the UE is actually an attacker who tries to run a denial-of-service attack against the network.

In both cases, the best course of action is to deny access to the UE in question.

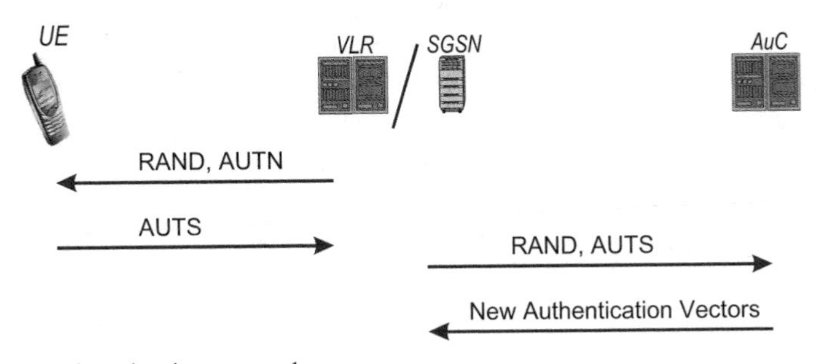

Figure 2.6 Resynchronization procedure
UE = User Equipment; VLR = Visitor Location Register; SGSN = Serving GPRS Support Node; AuC = Authentication Centre; RAND = random number; AUTN = authentication token; AUTS = authentication token in re-synchronization; GPRS = General Packet Radio Service

2.1.1.6 Illustrative flow chart of authentication

In Figure 2.7 we show a flow chart outlining the mutual authentication procedure, including the potential resynchronization phase. There also exists a procedure for reporting authentication failures from the VLR/SGSN to the Home Location Register (HLR) (see [9]). This procedure is not included in the flow chart.

2.1.2 Temporary identities

The permanent identity of the user in UMTS is IMSI (as is also the case in GSM). However, identification of the user in UTRAN (UMTS Terrestrial Radio Access Network) is in almost all cases effected by means of temporary identities: TMSI in the CS domain or P-TMSI in the PS domain. Confidentiality of user identity is thus protected (almost always) against passive eavesdroppers. Initial registration is, of course, the exception because a temporary identity cannot be used since the network does not yet know the permanent identity of the user. After that it is possible to use temporary identities.

The mechanism works as follows. Assume the user has already been identified in the SN by IMSI. Then the SN (VLR or SGSN) allocates a temporary identity (TMSI or P-TMSI) for the user and maintains an association between the permanent identity and the temporary identity. The latter only has local value and each VLR/SGSN simply takes care that it does not allocate the same TMSI/P-TMSI to two different users simultaneously. The allocated temporary identity is transferred to the user once encryption is turned on. This identity is then used in both uplink and downlink signalling until the network allocates a new TMSI (or P-TMSI). Paging, location update, attach and detach are examples of signalling that utilizes (P-)TMSI.

Allocation of a new temporary identity is acknowledged by the terminal, and then the old temporary identity is removed from the VLR (or SGSN). If allocation acknowledgement is not received by the VLR/SGSN it keeps both the old and new TMSIs and accept either of them in uplink signalling. In downlink signalling, IMSI must be used because the network does not know which temporary identity is currently stored in the terminal. In this case, VLR/SGSN tells the terminal to delete any stored TMSI/P-TMSI and a new reallocation follows.

However, one problem remains: how does the SN obtain the IMSI in the first place? Since the temporary identity only has local meaning, the identity of the local area has to be appended to it in order to obtain a unique identity for the user. This is resolved by appending the Location Area Identity (LAI) to the TMSI and the Routing Area Identity (RAI) to the P-TMSI.

If the UE enters a new area, then the association between IMSI and (P-)TMSI can be fetched from the old location or routing area if the new area knows its address

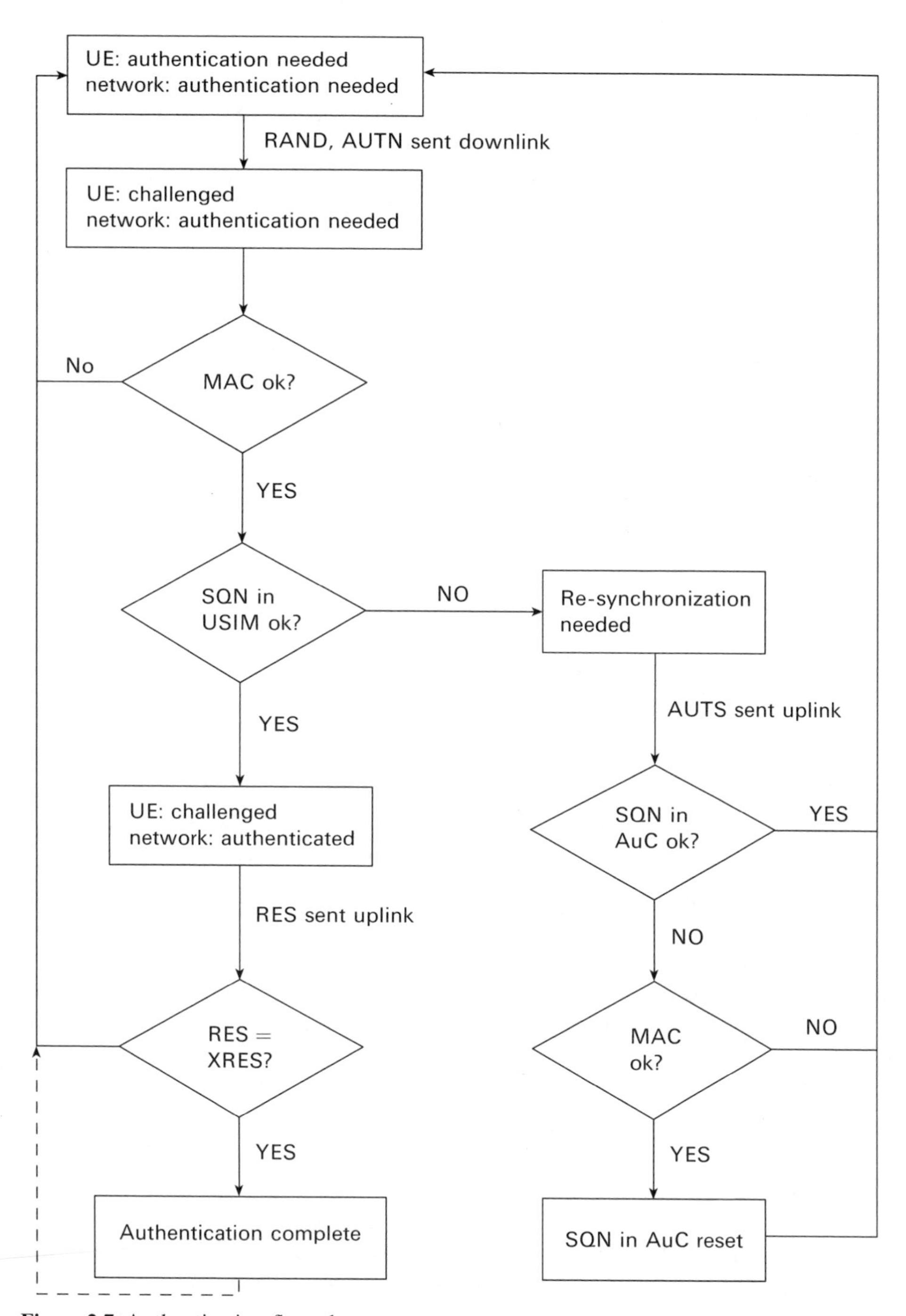

Figure 2.7 Authentication flow chart

UE = User Equipment; RAND = random number; AUTN = authentication token; MAC = Message Authentication Code; SQN = sequence number; USIM = Universal Subscriber Identity Module; AUTS = authentication token in re-synchronization; AuC = Authentication Centre; RES = user response; XRES = expected response

(based on LAI or RAI). At the same time, unused AVs can also be transferred from the old VLR/SGSN to the new VLR/SGSN (if there are any). If the address is not known or a connection to the old area cannot be established, then the IMSI must be requested from the UE.

There are certain places, such as airports, where lots of IMSIs may be transmitted over the radio interface as people switch on their mobile phones after the flight. This means that the people arriving can in principle be identified should an eavesdropper know their IMSIs. On the other hand, the ability to track people is also easier in such places (e.g., by observing who gets off which plane!).

Although the user identity confidentiality mechanism in UMTS does not give 100% protection, it offers a pretty good level of protection. Note that protection against an active attacker is not very good since the attacker may pretend to be a new SN and the user is likely to reveal his or her permanent identity. The mutual authentication mechanism does not help here since the user has to be identified before he or she can be authenticated.

Further details about handling temporary identities can be found in [29] and [2].

2.1.3 UTRAN encryption

Once the user and the network have authenticated each other they may begin secure communication. As described earlier, a CK is shared between the CN and the terminal after a successful authentication event. Before encryption can begin, the communicating parties also have to agree on the encryption algorithm. In a UMTS, implemented according to 3GPP Release 1999, only one algorithm is defined. At the time of writing, the specification process has begun with the remit of designing another encryption algorithm for fallback purposes.

It is in general a good security principle to take precautions against the potential situation where the cryptographic algorithm used in the system suddenly fails. Although there is typically a time gap between any first theoretical attack and widespread practical attacks, this time is not necessarily long enough to allow introduction of another algorithm. If two algorithms could be used at the same time, then if one of them fails the security of the system would not be jeopardized.

Encryption and decryption take place in the terminal and in the RNC on the network side, which means that the CK has to be transferred from the core network (CN) to the RAN. This is done in a specific Radio Access Network Application Protocol (RANAP) message, called the *security mode command*. After the RNC has obtained the CK, it can switch encryption on by sending a Radio Resource Control (RRC) security mode command to the terminal.

The UMTS encryption mechanism is based on a *stream cipher* concept as described in Figure 2.8. This means that plaintext data are added bit by bit to random-

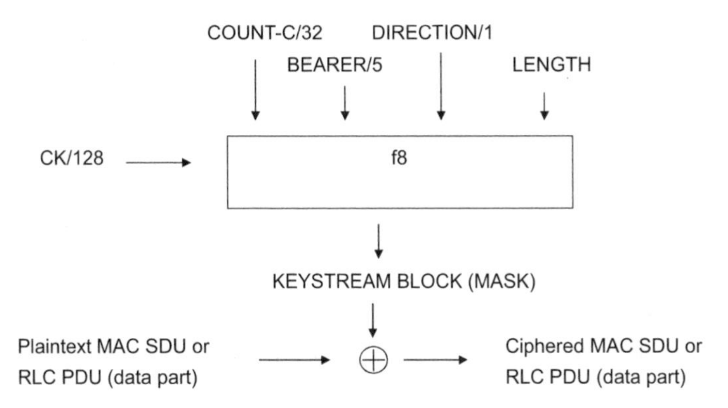

Figure 2.8 Stream cipher concept
CK = Cipher Key; MAC SDU = Medium Access Control Signalling Data Unit; RLC PDU = Radio Link Control Protocol Data Unit

looking mask data that are generated by the CK and a few other parameters. This type of encryption has the advantage that the mask data can be generated even before the plaintext is known, resulting in final encryption being a very fast bit operation. Decryption on the receiving side is done in exactly the same way, since adding the mask bits twice has the same result as adding zeros.

Because mask data do not depend on plaintext, there has to be another input parameter that changes every time a new mask is generated. Otherwise, the same mask would protect two different plaintexts, say P_1 and P_2, resulting in the following unwanted phenomenon: if we add P_1 to P_2 bit by bit and do the same to their encrypted counterparts, then the resultant bit string is exactly the same in both cases. This is a consequence of the fact that two identical masks cancel each other during bit-by-bit addition. Therefore, any attacker who eavesdrops the corresponding encrypted messages on the radio interface would know the bit-by-bit sum of P_1 and P_2. So, if two bit strings of meaningful data are added to each other bit by bit, their content could be discovered from the resultant bit string, which means encryption of the two messages P_1 and P_2 would be broken. The example below illustrates how effective this kind of break is.

2.1.3.1 Example: breaking encryption when mask is reused

Plaintext always has some structure. It is not just random data, it contains some *redundancy*. In our example we assume the plaintexts in question are in English. When coded into ASCII bit strings this assumption implies huge redundancy for these bit strings: most ASCII codes never appear and some appear very frequently. For illustrative purposes, however, let us assume a simplified coding for this example: we use only capital letters from A to Z plus space in between words

Table 2.2 Simplified coding for the English alphabet

A	B	C	D	E	F	G	H	I	J	K	L	M	N
0	1	2	3	4	5	6	7	8	9	10	11	12	13
O	**P**	**Q**	**R**	**S**	**T**	**U**	**V**	**W**	**X**	**Y**	**Z**	**Space**	
14	15	16	17	18	19	20	21	22	23	24	25	26	

(i.e., no punctuation). Letter A is coded as 0. Similarly, B is coded as 1, C as 2, etc. Finally, Z is coded as 25 and space is coded as 26. The full coding is given in Table 2.2.

Encryption is effected as follows. A list of random integers from the interval between 0 and 26 is created and serves as a mask. Each plaintext letter, or more precisely, the number corresponding to it, is added to an entry in the random integer list in a modular fashion (i.e., the numbers 27 and 0 are considered to be equal). Also, for each number that exceeds 27, multiples of 27 are subtracted until the result is between 0 and 26 (e.g., $14 + 21 = 35$, but when 27 is subtracted, the result is 8). For example, assume the plaintext is CAT while the mask is (3, 17, 12). Then the encoded plaintext is (2, 0, 19) and the ciphertext is (5, 17, 4).

This encryption provides perfect security as long as the mask is truly random, is not known to the attacker and is *used only once*. Indeed, any three-letter plaintext could be transformed to the same ciphertext (5, 17, 4) with a suitable mask. MOM encodes to (12, 14, 12) and the mask producing our ciphertext would be (20, 3, 19). Similarly, XYZ is potential plaintext if the mask happens to be (9, 20, 6).

Let us now assume that the same mask has been used to encrypt two different (but extremely) short English texts. Let us try and discover the contents of both those texts. The ciphertexts are the following:

1. (21, 12, 22, 25, 21, 15, 6)
2. (6, 15, 9, 20, 13, 0, 1)

Let us start our analysis with the first two letters:

1. 21 12
2. 06 15

There exist various statistics about the frequencies of letters in average English text as well as statistics about the frequencies pairs of letters (*digrams*) (e.g., [52]). For instance, the 15 most common digrams cover almost 30% of all cases. Thus, we could try to jump-start the analysis by testing the most frequent digrams for the first two letters of the unknown plaintexts.

The most frequent digram in English is TH, encoded as 19 07. If plaintext 1 began with TH, then the mask would be 02 05 and plaintext 2 would begin with 04 10, decoded as EK. This is certainly possible in principle but not very promising, as EK is not among the top 100 commonest digrams and very few English words begin with it.

Next let us try TH for the beginning of plaintext 2: in this case the mask would be 14 08 and plaintext 1 would start with 07 04 (i.e., HE). This is more promising, as HE is one of the most frequent digrams.

The most probable continuation for plaintext 2 is THE + space, encoded 19 07 04 26. This yields a mask 14 08 05 21 and plaintext 1 would be encoded 07 04 17 04, corresponding to HERE. We still seem to be on the right track.

Plaintext 1 continues with a space, hence the next element in the mask is 22 and plaintext 2 continues with 18 (i.e., plaintext 2 is THE S??). The best tactics to adopt now seem to be testing common three-letter words that begin with S. Plaintext 2 = THE SEA would imply plaintext 1 = HERE TF, no good. THE SKY would imply HERE RD, no better. Finally, plaintext 2 = THE SUN implies plaintext 1 = HERE IS, and a very probable solution is found: "Here is ... the sun".

Of course, with an automated procedure we could do much more, but this analysis at least gives an idea about how the analysis of longer texts could be easily done.

2.1.3.2 Encryption parameters

UTRAN encryption occurs in either the Medium Access Control (MAC) layer or in the Radio Link Control (RLC) layer. In both cases, there is a counter that changes for each Protocol Data Unit (PDU). In the MAC this is the Connection Frame Number (CFN) and in the RLC it is a specific RLC sequence number (RLC-SN). If these counters were used as input for mask generation the problem explained in the previous paragraph would still occur since these counters wrap around very quickly. This is why a longer counter called a Hyper Frame Number (HFN) is introduced. It is incremented whenever the short counter (CFN in the MAC case and RLC-SN in the RLC case) wraps around. The combination of HFN and the shorter counter is called COUNT-C and is used as ever-changing input to mask generation inside the encryption mechanism.

In principle, the longer counter HFN could also eventually wrap around. Fortunately, it is reset to zero whenever a new key is generated during the AKA procedure. Authentication events are in practice frequent enough to rule out the possibility of HFN wrap-around.

The radio bearer identity BEARER is also needed as an input to the encryption algorithm since the counters for different radio bearers are maintained independently of each other. If the input BEARER was not in use, then this would again lead to a situation where the same set of input parameters are fed into the algorithm and the same mask would be produced more than once. Consequently, the problem outlined in the example above would occur and the messages (this time in different radio bearers) encrypted with the same mask would be exposed to the attacker.

The DIRECTION parameter indicates whether we encrypt uplink or downlink traffic. The LENGTH parameter indicates the length of data to be encrypted. Note that the value of LENGTH only affects the *number* of bits in the mask bit stream, it does not have any effect on the bits themselves in the generated stream.

The core of the encryption mechanism is the mask generation algorithm, which is denoted as function f8. The specification is publicly available as 3GPP TS 35.201 [19] and is based on a novel *block cipher* called KASUMI (for which there is another 3GPP specification TS 35.202 [20]). This block cipher transforms 64-bit input to 64-bit output. The transformation is controlled by the 128-bit CK. If CK is not known, then there are no efficient algorithms to compute the output from the input or vice versa. In principle, transformation can be done if:

- all possible keys are tried until the correct one is found; or
- an enormous table of all 2^{64} input–output pairs is assembled.

However, both approaches are impossible in practice. These algorithms are presented in detail in Chapters 6 and 7.

It is possible for authentication not to be carried out at the beginning of the connection. In this case the previous CK is used for encryption. The key is stored in the USIM between the connections. The START parameter, which consists of the most significant part of the greatest HFN used so far, is also stored in the USIM. For the next connection, the stored value is incremented by 2 and used as the starting value for the most significant part of HFN. There is also a constant parameter in USIM, called THRESHOLD, that may be used to restrict the maximal lifetime of the keys CK and IK. Whenever START reaches THRESHOLD, generation of new keys is forced by the UE (i.e., the UE informs the network it has no valid keys).

2.1.3.3 UTRAN protocol structure

As the encryption mechanism is built into radio network protocols, we will discuss these protocols in this chapter. The protocols in the RAN in UMTS are divided into three layers:

1. physical layer;
2. link layer;
3. network layer.

This division follows the classical OSI (Open Systems Interconnection) model. Furthermore, layer 2 is divided into several sublayers:

- MAC;
- RLC;
- Packet Data Convergence Protocol (PDCP);
- Broadcast/Multicast Control (BMC).

The physical layer and the MAC support both user (U-) plane and control (C-) plane traffic in an essentially similar manner. Both the PDCP and the BMC only exist in the U-plane, while the RLC and layer 3 are divided into U-plane and C-plane.

Layer 3 is also divided into several sublayers. However, only the lowest sublayer, the RRC, terminates in the UTRAN in the RNC. Higher sublayers terminate in the CN. The RRC protocol exists only in the C-plane. There are two kinds of control messages transported over the radio interface: radio-specific messages generated by RRC and NAS (Non Access Stratum) control messages generated by higher layers. The NAS control traffic includes the Mobility Management (MM) and Call Control (CC) protocols. The RRC sublayer also provides interlayer communication with all lower layers, thus taking care of their configuration.

The services provided by each of the UTRAN layers to higher layers are summarized in the following.

2.1.3.3.1 Physical layer

Physical layer services convert *physical* radio channels to *transport* channels. These can be characterized as describing *how* the data are transferred rather than *what* data are transferred. Layer 1 services include error detection and correction, frequency and time synchronization, multiplexing of transport channels, interleaving, modulation, power control, measurements and execution of soft handovers, among others. The transport channels are divided into two main categories:

- common channels—if only one particular UE (User Equipment) needs to be addressed, inband signalling is used;
- dedicated channels (DCH)—the whole channel is reserved for one particular user.

Common channels include the Random Access Channel (RACH) for transmitting short uplink messages (e.g., for initial access), the Forward Access Channel (FACH) for short downlink messages, the Paging Channel (PCH) and the Broadcast Channel (BCH), among others. In GSM, encryption is also done in the physical layer.

As the physical layer terminates at the BS, an important target for improved security in UMTS (compared with GSM) was to move the termination point of encryption further back into the network. For this reason, encryption is not done in the physical layer in UMTS.

2.1.3.3.2 MAC

The MAC layer converts transport channels into *logical channels*, which are characterized by *what* kinds of data are transferred. The main division of logical channels is the following:

- traffic channels—for U-plane information;
- control channels—for C-plane information.

Logical channels include the broadcast control channel, paging control channel, common control channel (CCCH), dedicated control channel (DCCH), common traffic channels and dedicated traffic channels. These logical channels are mapped into transport channels (e.g., the broadcast control channel can be mapped into either BCH or FACH and the dedicated traffic channel can be mapped into RACH, FACH, DCH, etc.).

The MAC layer contains, among others, the following functions:

- mapping logical channels into transport channels;
- choosing an appropriate transport format for each transport channel;
- identification of an addressed UE in common channels (this is the inband signalling referred to above);
- multiplexing of upper layer PDUs;
- traffic volume measurement.

The MAC layer also performs encryption in transparent RLC mode (e.g., in case of CS speech traffic). In this case the part that is encrypted is the MAC SDU (Signalling Data Unit) but the MAC header is not. The counter CFN consists of the least significant part of the encryption counter COUNT-C.

It is possible that several MAC PDUs are transmitted during the same Transmission Time Interval (TTI). In this case ciphering is not initialized in the middle of the TTI. Instead, the input parameter COUNT-C for the whole TTI is obtained from the CFN of the first radio frame in the TTI. Then a long mask bit stream is generated and used to encrypt all the radio frames in the TTI.

2.1.3.3.3 RLC

The RLC layer provides the following services to the upper layers:

- transparent data transfer—upper layer PDUs are transmitted without any additional protocol information except possibly segmentation/reassembly of them;
- unacknowledged data transfer—upper layer PDUs are transmitted without guarantees of delivery, but with detection of transmission errors;
- acknowledged data transfer—upper layer PDUs are transmitted with guaranteed delivery, potential retransmissions are used for error-free delivery and double transmissions are also detected;
- maintenance of Quality of Service (QoS) as defined by upper layers;
- notification of irrecoverable errors to upper layers.

The most important RLC functions are: segmentation and reassembly of upper layer PDUs; concatenation of the first segment of an RLC SDU with the last segment of the previous RLC SDU into the same RLC PDU, adding padding bits in case no concatenation is possible; data transfer (transparent, unacknowledged or acknowledged); error correction; in-sequence delivery of upper layer PDUs; duplicate detection; RLC SQN check (in unacknowledged mode to provide the possibility of detecting errors when RLC PDUs are reassembled into RLC SDUs); protocol error detection and recovery.

The RLC layer also provides encryption in unacknowledged and acknowledged RLC modes, when ciphering is applied to the whole RLC PDU except the PDU header. The header consists of a SQN (7 bits) and an extension bit (making one octet) in the UM (Unacknowledged Mode) case, and of a SQN (12 bits) and 4 other bits (making two octets) in the AM (Acknowledged Mode) case. In the former case the extension bit of the header indicates whether a *length indicator* follows or the data. In the AM case the 4 bits that are included in the header in addition to the SQN indicate:

- whether the PDU contains control information or data;
- whether a status report (status PDU) is requested from the receiver;
- whether the length indicator follows or the data.

The structure of lower-layer data units is illustrated Figures 2.9–2.12.

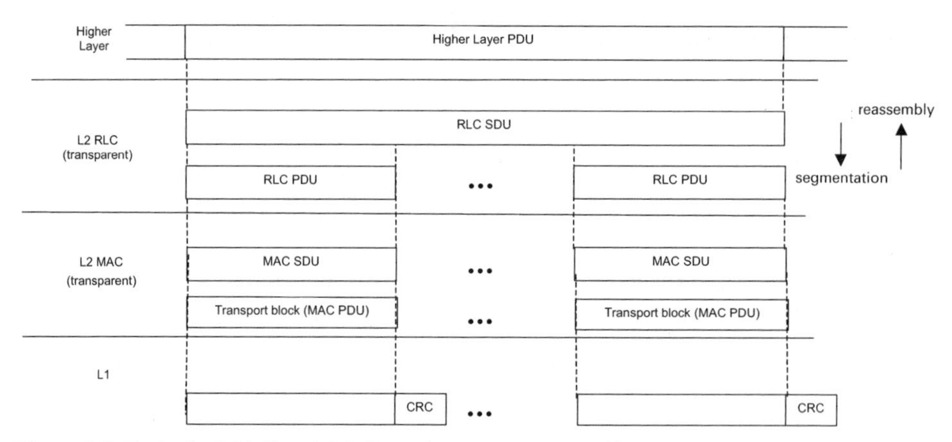

Figure 2.9 Both the MAC and RLC are in transparent mode
MAC = Medium Access Control; RLC = Radio Link Control; PDU = Protocol Data Unit; SDU = Signalling Data Unit; CRC = Cyclic Redundancy Check

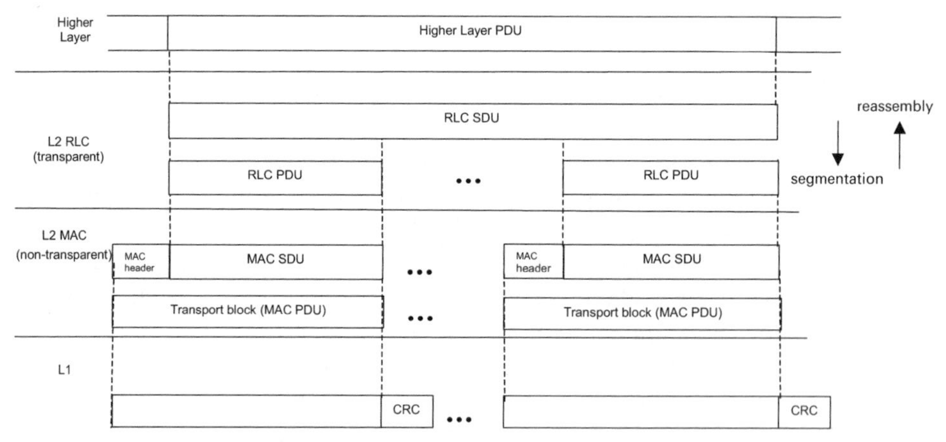

Figure 2.10 Transparent RLC mode and non-transparent MAC mode
RLC = Radio Link Control; MAC = Medium Access Control; PDU = Protocol Data Unit; SDU = Signalling Data Unit; CRC = Cyclic Redundancy Check

2.1.3.3.4 PDCP

The PDCP provides header compression/decompression of IP (Internet Protocol) traffic (e.g. for TCP (Transmission Control Protocol) or IP headers), among other things.

2.1.3.3.5 BMC

The BMC provides transmission and scheduling of BMC messages, and storage and delivery of cell broadcast messages.

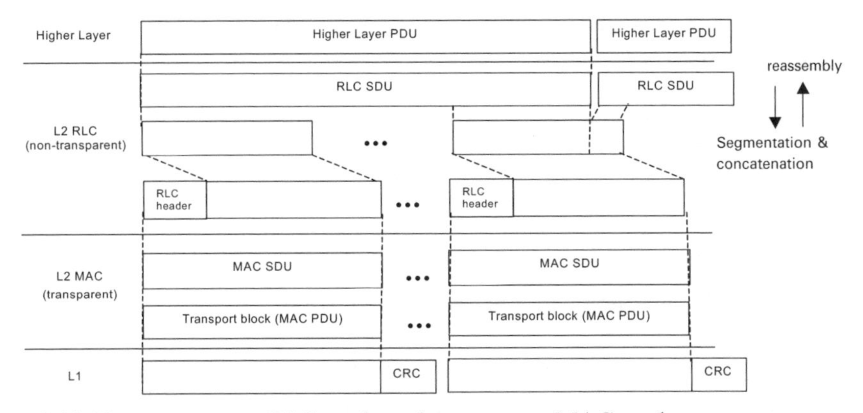

Figure 2.11 Non-transparent RLC mode and transparent MAC mode
RLC = Radio Link Control; MAC = Medium Access Control; PDU = Protocol Data Unit; SDU = Signalling Data Unit; CRC = Cyclic Redundancy Check

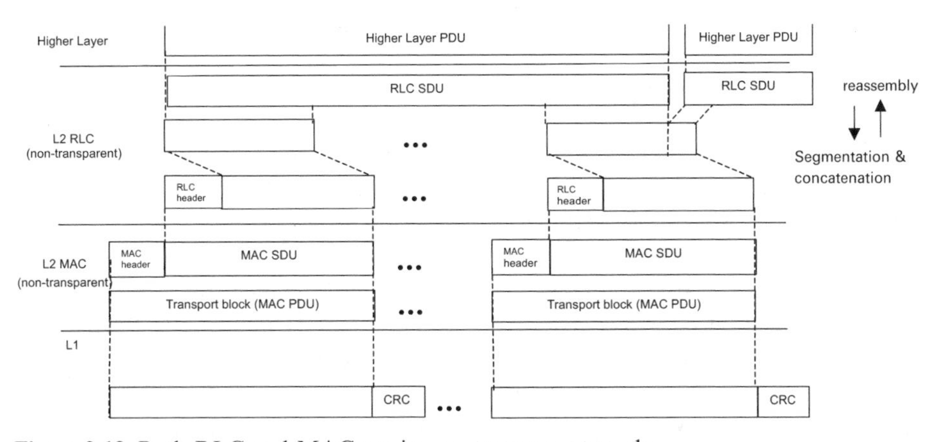

Figure 2.12 Both RLC and MAC are in non-transparent mode
RLC = Radio Link Control; MAC = Medium Access Control; PDU = Protocol Data Unit; SDU = Signalling Data Unit; CRC = Cyclic Redundancy Check

2.1.3.3.6 RRC

The RRC provides such functions as:

- broadcast of both NAS and Access Stratum (AS) information—for NAS information (e.g., general system-level information), the RRC provides scheduling, segmentation and repetition (AS information is typically cell-specific information about the radio environment);
- establishment, re-establishment, maintenance and release of RRC connections between the UE and RNC;

- establishment, reconfiguration and release of (U-plane) radio bearers requested by upper layers;
- RRC connection mobility functions, such as handovers, preparations for handovers to GSM, cell reselection;
- paging and notification requested by upper layers;
- control of requested QoS (appropriate radio resources have to be provided);
- control of UE measurements and related reporting.

The RRC also provides encryption control (i.e., it decides whether encryption is on or off between the UE and RNC) as well as executing *integrity protection* of both RRC-level signalling and higher layer signalling in the form of message authentication codes (MAC-I, the "I" stands for integrity of signalling data).

2.1.3.4 UE modes and identification

Inside UTRAN the UE can be in two different modes: *idle* or *connected.* After power has been switched on, the UE is in idle mode. When an RRC connection is established between the UE and RNC, the UE enters the connected mode and when the RRC connection is released the UE returns to idle mode.

In idle mode, the UE can only be identified by CN-level identities (i.e., by IMSI, TMSI or P-TMSI). In connected mode, it is also possible to use a UTRAN-level identity called Radio Network Temporary Identity (RNTI).

A necessary requirement for authentication is that the UE be identified first; hence, used identities play an important role in the overall security architecture of the system. Authentication is always done in NAS-level signalling and therefore is tied to IMSI, TMSI or P-TMSI. Integrity protection can be defined as *authentication of individual messages* and this type of authentication may be based on RNTI as well. As a consequence, the network must maintain the connection between NAS-level identities and RNTIs.

2.1.4 Integrity protection of RRC signalling

The purpose of integrity protection is to authenticate individual control messages. This is important, since separate authentication procedures only give assurance of the identities of the communicating parties at the time of the authentication. This leaves the door open for an attacker called "the man in the middle" to act as a simple relay and deliver all messages in their correct form until the authentication procedure is completely executed. After that, the man in the middle may begin to manipulate

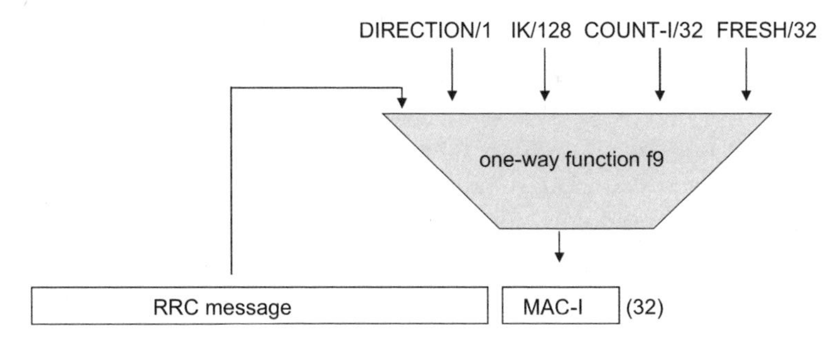

Figure 2.13 Message authentication code
IK = Integrity Key; RRC = Radio Resource Control; MAC-I = Message Authentication Code

messages freely. However, if messages are protected individually, deliberate manipulation of messages can be observed and false messages can be discarded.

Integrity protection is implemented at the RRC layer (i.e., between the terminal and RNC), just like the case for encryption. The IK is generated during the AKA procedure, again in the same way as the CK is generated. The IK is transferred to the RNC together with the CK in *security mode command.*

The integrity protection mechanism is based on the concept of a *message authentication code*, which is a one-way function controlled by the secret IK. The function is denoted by f9 and its output is MAC-I: a 32-bit, random-looking bit string. On the sending side, the MAC-I is computed and is appended to each RRC message. On the receiving side, the MAC-I is also computed and the result of the computation is checked to ensure it equals the bit string appended to the message. Any change in any of the input parameters affects the MAC-I in an unpredictable way.

The function f9 is depicted in Figure 2.13. Its inputs are IK, the RRC message itself, a counter COUNT-I, direction bit (uplink/downlink) and a random bit string called FRESH. The COUNT-I parameter resembles the corresponding counter for encryption. Its most significant part is an HFN that consists of 28 bits in this case, and the four least significant bits contain the RRC sequence number. Altogether, COUNT-I protects against replay of earlier control messages by guaranteeing that the set of values for input parameters is different for each run of the integrity protection function f9.

The algorithm for integrity protection is based on the same core function as encryption. Indeed, the KASUMI block cipher is used in a special mode to create a message authentication code function. A detailed description of the first 3GPP integrity protection algorithm is given in Section 6.8. At the time of writing, specification work has begun on defining another integrity algorithm for fallback purposes.

The FRESH parameter is chosen by the RNC and transmitted to the UE. It is needed to protect the network against a maliciously-chosen start value for COUNT-I. Indeed, the most significant part of HFN is stored in the USIM between connections. An attacker could masquerade as the USIM and send a false value to the network, forcing the starting value of HFN to be too small. If the authentication procedure is not run and the old IK is brought into use, this would create a chance for the attacker to replay RRC signalling messages from earlier connections with recorded MAC-I values were the FRESH parameter not involved. By choosing FRESH randomly, the RNC is protected against such a replay attack (i.e., based on recording of earlier connections). As already explained, the ever-increasing counter COUNT-I protects against replay attacks that are based on recording during the *same* connection because FRESH stays constant over a single connection. From the terminal's point of view, it is still essential that the value COUNT-I never repeats itself even between different connections, because a false network could send an old FRESH value to the UE in order to try a replay attack in the downlink direction.

Note that radio bearer identity is not used as an input parameter for the integrity algorithm, although it is an input parameter for the encryption algorithm. Because there are also several parallel radio bearers for the control plane, this seems to leave room for possible replay of control messages that were recorded within the same RRC connection but on a different radio bearer. There is a historical reason for this state of affairs: at the time of freezing requirements for integrity protection algorithm design work, the specification for UTRAN contained only one signalling radio bearer.

Instead of changing the algorithm structure retrospectively, the following procedure was introduced in the integrity protection mechanism to remove the security hole. Radio bearer identity is always appended to the message when the message authentication code is calculated, although it is not transmitted with the message. So, not only does the radio bearer identity have an effect on the MAC-I value, we also have protection against replay attacks based on recordings from different radio bearers.

Clearly, there are RRC control messages whose integrity cannot be protected by the mechanism. Indeed, messages sent before the IK is in place cannot be protected. A typical example is the *RRC connection request* message sent from the UE. The following list contains all messages that are not integrity-protected:

- handover to UTRAN complete;
- paging type 1;
- push capacity request;
- physical shared channel allocation;
- RRC connection request;
- RRC connection set-up;

- RRC connection set-up complete;
- RRC connection reject;
- RRC connection release (CCCH (Common Control Channel) only);
- system information (broadcast information);
- system information change indication;
- transport format combination control (TM (Transparent Mode) DCCH only).

2.1.4.1 Periodic local authentication

The integrity protection mechanism in UTRAN is not applied for the U-plane for performance reasons. However, there is a specific (integrity-protected) control plane procedure that is used for *periodic local authentication*. As a result of this procedure, the amount of data sent during the RRC connection is checked. Hence, the *volume* of transmitted user data is integrity-protected and at the same time, the procedure provides local entity authentication.

Periodic local authentication is initiated by the RNC and triggered by some COUNT-C value that reaches a *critical value* (e.g., a certain bit in the HFN changes). Then the RNC sends a *counter check* message that contains the most significant part from each COUNT-C, corresponding to each active radio bearer. The UE compares the sent values with the most significant parts of its own COUNT-C values. All differences are reported back in a *counter check response* message. If the response message does not contain any values, then the procedure ends. If there are differences, the RNC may release the connection (in the event the differences cannot be accepted). The procedure is depicted in Figure 2.14.

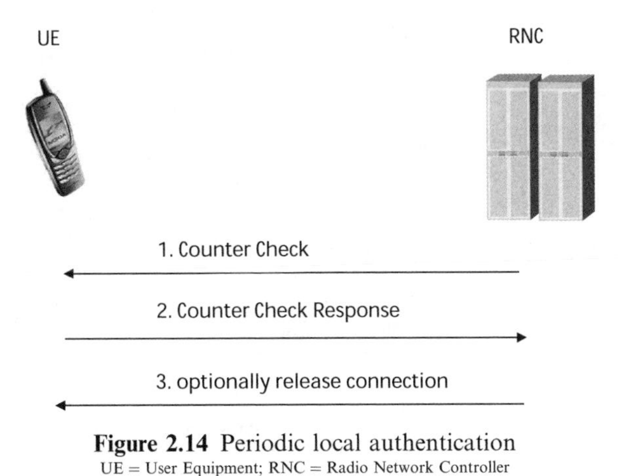

Figure 2.14 Periodic local authentication
UE = User Equipment; RNC = Radio Network Controller

Periodic local authentication gives protection against an attacker who tries to insert or delete data packets during uplink or downlink. The protection is especially important in case encryption is not in use. Note that in this case both the UE and the RNC need to maintain COUNT-C values despite the fact they are not used for encryption.

It is possible that the attacker could try and insert and delete the same number of packets in order to keep COUNT-C values synchronized. The legitimate user cannot stop this type of attack unless he or she notices a drop in service level.

2.1.4.2 Threats against UTRAN signalling

In this section we elaborate a bit more on the different types of threats that UTRAN signalling is exposed to. A simplified picture of the UTRAN link layers and network nodes involved is depicted in Figure 2.15.

The main signalling flows take place between the UE and the RNC. Signalling messages are exchanged at all three layers: RRC, RLC and MAC. The most important and sensitive are those on the RRC layer and their integrity is protected. Encryption provides protection for signalling in RLC and MAC.

In the rest of this section we give examples of the threats to signalling in each layer (the functions performed at each layer were listed in Section 2.1.3.3.

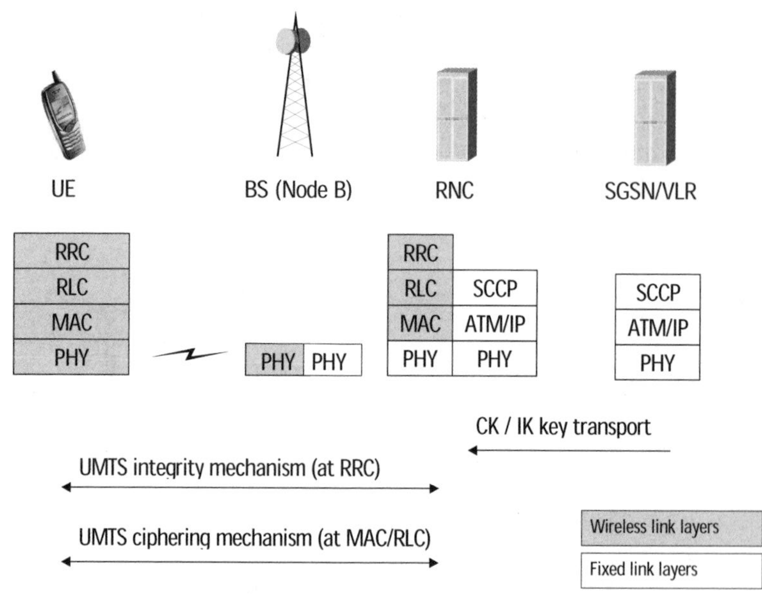

Figure 2.15 Protocols in UTRAN
UTRAN = UMTS Terrestrial Radio Access Network; UE = User Equipment; RRC = Radio Resource Control; RLC = Radio Link Control; MAC = Medium Access Control; PHY = physical layer; BS = Base Station; RNC = Radio Network Controller; SCCP = Signalling Connection Control Part; ATM = Asynchronous Transfer Mode; IP = Internet Protocol; CK = Cipher Key; IK = Integrity Key

- *Threats to MAC layer signalling*—
  - identification information sent by the UE over common transport channels can be tampered with;
  - access service class selection for RACH and CPCH (Common Packet Channel) transmission can be tampered with.
- *Threats to RLC functions*—
  - if RLC PDU (Protocol Data Units) headers are tampered with, duplicate detection is made impossible;
  - attackers can tamper with flow control messages and in this manner disturb the traffic flow and deteriorate the QoS of the victim UE;
  - attackers can tamper with SQNs, thus preventing the detection of corrupt RLC SDUs (Signalling Data Units).
- *Threats to RRC functions*—some of the threats to the RRC are addressed by the integrity protection mechanism, but as mentioned earlier in this section not all RRC messages can be protected—
  - broadcast of information provided by the CN cannot be integrity-protected;
  - broadcast of (typically cell-specific) information provided by the AS cannot be integrity-protected;
  - an attacker UE can try to hijack the connection by tampering with RRC connection re-establishment requests sent by the victim UE;
  - significant damage can be caused by tampering measurements made and sent by the victim UE;
  - paging information and target addresses can be tampered with;
  - measurement information sent by the victim UE can be tampered with;
  - idle mode measurements sent by the victim UE can be tampered with, thus forcing it to a non-suitable cell;
  - non-optimal performance can be caused by tampering with configuration messages sent by the RRC to the victim UE.

2.1.5 Set-up of UTRAN security mechanisms

The use of encryption and integrity protection is vital for the security of UTRAN. To guarantee that mechanisms cannot be bypassed, it is important to define exactly how they are turned on after communication has been established. As mentioned earlier, the use of encryption is not mandatory in the system, but there must not be a way of avoiding the application of integrity protection.

In this section we describe how encryption and integrity protection are activated at the time of connection set-up.

2.1.5.1 Negotiation of the algorithms

Assume the UE wants to establish a connection with the network. First, a *classmark* that indicates the capabilities of the UE is sent to the network. These capabilities always include support for different encryption and integrity algorithms. Because this information is transferred at the very beginning of the connection, though the purpose of the information is to establish security later, there cannot be any protection for the transmitted classmark at this point. This problem is addressed by rechecking classmark information at a later stage of the security set-up procedure. Based on the received classmark, the network decides which algorithms to use:

- if there are no integrity protection algorithms in common, then the connection is shut down immediately;

- if there are no encryption algorithms in common, then the network may establish the connection without encryption.

All UEs compliant with 3GPP Release 1999 standards must support the integrity algorithm UIA1, hence the first case above should never occur.

Because the UE can establish connections in CS and PS domains independently of each other, in principle it could be possible for the network to choose different algorithms for different domains. However, this is not permitted. The reason for this is that security algorithms are implemented in the RNC, which is a common element of both CN domains. Therefore, using different algorithms for different domains would be an unnecessary burden for the RNC (and for the UE as well).

2.1.5.2 Existing parameters in USIM

When a new connection is established, some parameters are inherited from the previous connection (failing that, some links to the previous connection are needed).

The UE has stored the security keys that have been used for both domains (up to four keys in total). The UE has also stored the value of START for both CS and PS domains to the USIM. At the same time whether either of these values have reached the maximal allowed value, called THRESHOLD, has been checked. The latter is configured to the USIM and provides a means of limiting security key lifetimes. If START has reached THRESHOLD for a CN domain, then the CK and IK for this domain are deleted from the USIM and START is set to be equal to THRESHOLD.

At the beginning of a new connection, START values and security keys are read from the USIM. The Key Set Identifier (KSI) is associated with a pair of security keys: the CK and IK that were generated during the same run of an AKA procedure. The KSI consists of three bits: the value "111" is reserved just in case there are no valid keys in the USIM. The value of KSI wraps around fairly often, every seventh time, but a period of this length is enough to remove the risk of ambiguity in practice. The values of START and KSI are transmitted to the network as part of the first messages as soon as the connection is made.

2.1.5.3 Security mode set-up procedure

We now describe those steps that are followed when integrity protection (and possibly encryption) are turned on. Integrity protection is not turned on:

1. if the connection is only for periodic location registration (without any change in registration information);
2. if the connection is only for indicating deactivation from the UE;
3. if authentication fails and, therefore, connection is immediately shut down;
4. if the connection is for an *emergency call* and there is neither a USIM nor a SIM (Subscriber Identity Module) in the UE.

Figure 2.16 describes the messages and security-relevant information elements that are transferred between the UE, RNC and SGSN/VLR during the set-up procedure.

Note that the procedure explicitly defines which message is the first to be integrity-protected at both uplink and downlink. On the other hand, this is not the case for encryption. At uplink the first encrypted message is the first message sent after the *security mode complete* message has been sent. At downlink the first encrypted message is the one sent after the *RANAP security mode complete* message has been sent to the CN. Because there may be messages in different layers waiting to be sent at the same time, it is not easy to decide which message should be the very first to be encrypted. For this purpose, a specific *ciphering activation time* parameter is exchanged between the UE and the RNC.

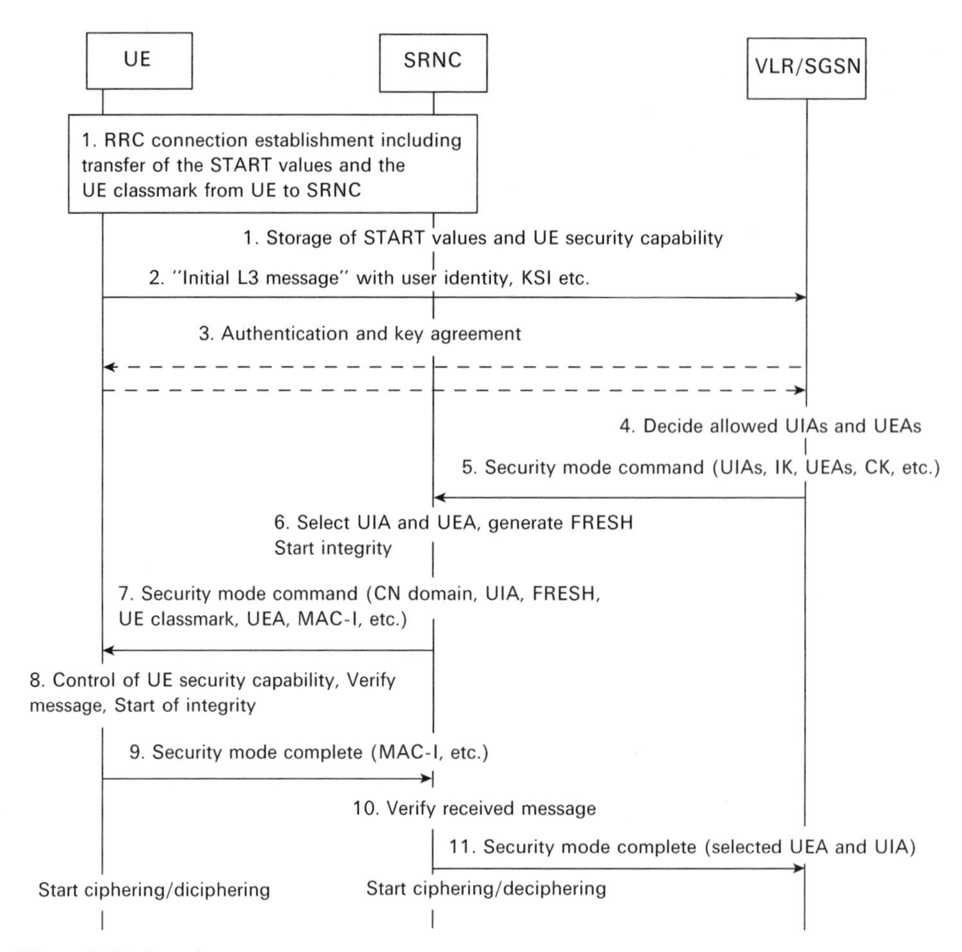

Figure 2.16 Security set-up
SRNC = Serving RNC; RNC = Radio Network Controller; VLR/SGSN = Visitor Location Register/Serving GPRS Support Node; GPRS = General Packet Radio Service; RRC = Radio Resource Control; UE = User Equipment; KSI = Key Set Identifier; UIA = UMTS Integrity Algorithm; UEA = UMTS Encryption Algorithm; IK = Integrity Key; CK = Cipher Key; CN = Core Network; MAC-I = Message Authentication Code

2.1.5.4 Security parameters for a new connection

If the AKA procedure is not carried out during connection establishment, then "old" security keys are used for the new connection. The COUNT-C and COUNT-I counter parameters are also initialized with the START value and the HFN is first initialized using START for the 20 most significant bits. The remaining HFN bits are set to 0. Note that different layers use HFNs of different lengths. The remaining COUNT bits are obtained from the layer-specific counters that are used for other purposes, in addition to security.

If the AKA procedure is carried out, then newly generated keys are used and START values are set to 0 at commencement of protection.

During an ongoing connection, START values are maintained in the ME: they consist of the greatest number obtained when comparing the 20 most significant bits of each COUNT value in each radio bearer that is in use for the CN domain in question.

In the U-plane, both the CS domain and PS domain use their own keys for protection algorithms, but the same *signalling* radio bearers are used for both domains. This implies that these signalling bearers have to use shared keys as well. Following general security principles, the keys generated most recently are used, regardless of whether they are for the CS domain or for PS domain. As a consequence, it is possible that protection keys may need to be changed for ongoing signalling connection.

2.1.6 Summary of access security in the CS and PS domains

We conclude Section 2.1 by presenting a schematic overview of the most important access security mechanisms and their relationships with each other. For the sake of clarity, many parameters are not shown in Figure 2.17 (e.g., HFN and FRESH are important parameters that are transmitted between different elements, but they are omitted from the figure).

2.2 Interworking with GSM

The UMTS CN is a straight evolution from that of GSM. The radio interfaces are completely different in both systems, but the early terminals still support both, allowing roaming from one system to another and, furthermore, handovers between the systems. As the security features in the two systems are different, it is not an easy task to define how security is managed during interoperation.

A smooth transition is needed from a pure GSM network to a mixed network that has wide area GSM coverage, enhanced by WCDMA islands. To enable this, the decision was taken that access to UTRAN would be possible with old SIM cards. Indeed, a user can use a 3G terminal without the need to change his or her smart card. The downside is a lower level of security: when a SIM card is used to access UTRAN, no authentication of the network is possible, because the card only provides 64 bits of key material (in the form of K_c) per authentication, while in the UTRAN side two 128-bit keys are needed. For this purpose, the 64-bit key K_c is *expanded* into 256 bits by using specific *conversion functions*. This procedure makes it possible to apply encryption and integrity protection in UTRAN when SIM cards

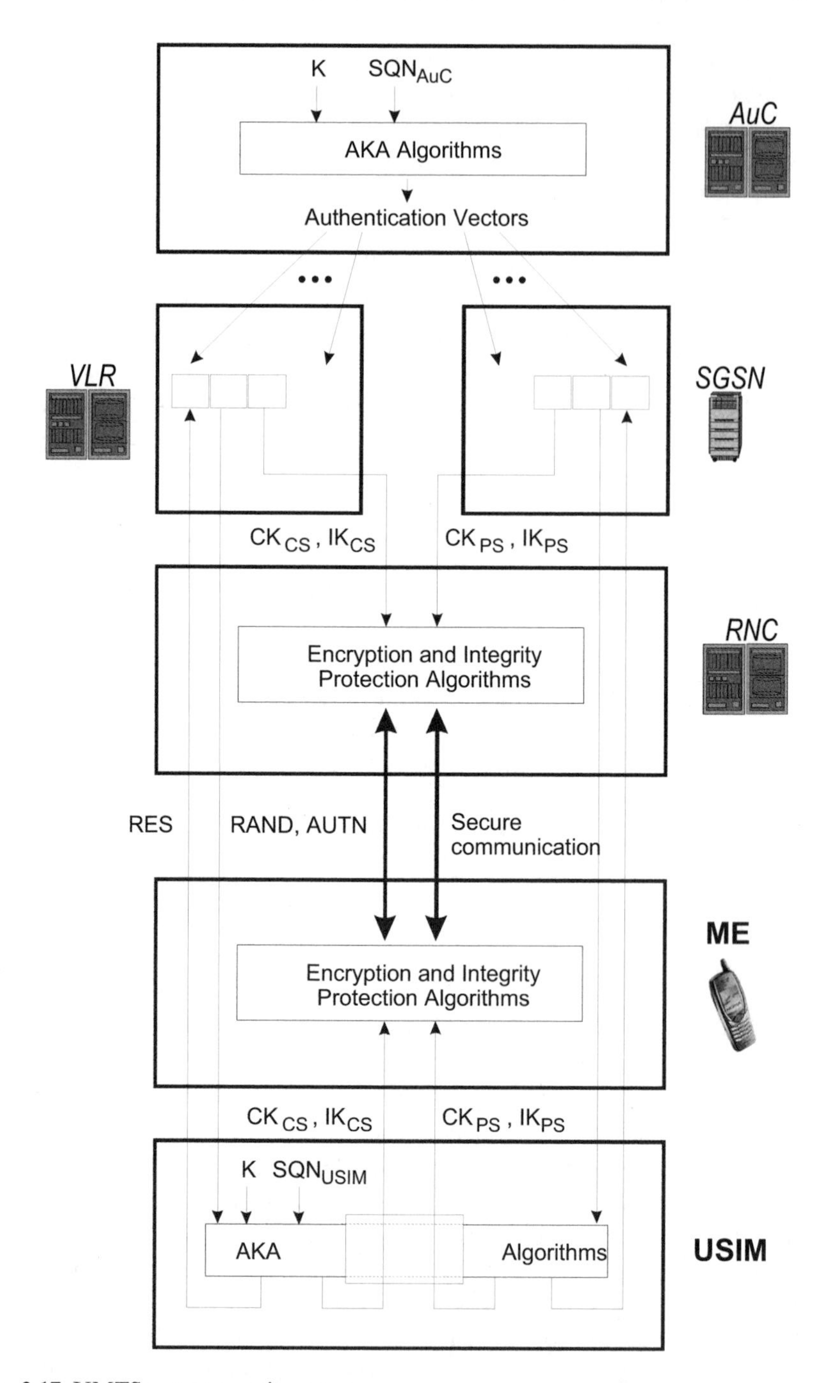

Figure 2.17 UMTS access security summary
SQN = sequence number; AuC = Authentication Centre; AKA = Authentication and Key Agreement; VLR = Visitor Location Register; SGSN = Serving GPRS Support Node; GPRS = General Packet Radio Service; CK = Cipher Key; CS = Circuit Switched; IK = Integrity Key; PS = Packet Switched; RNC = Radio Network Controller; RES = user response; RAND = random number; AUTN = authentication token; ME = Mobile Equipment; USIM = Universal Subscriber Identity Module

are used. However, the resulting security level can only be comparable with that of GSM because conversion functions only make keys longer *nominally*.

Another instance of interworking occurs when a 3G subscriber with a proper USIM needs to gain access outside WCDMA coverage. We then need to *compress* the longer keys provided by the USIM to 64 bits in order to use GSM encryption. Conversion functions are described in Section 8.8.

2.2.1 Interworking scenarios

In 3GPP technical report TR 31.900 [8], all possible interworking scenarios in a mixed 2G/3G environment are systematically studied. There are five basic entities in the system: the security module, terminal, radio network, serving CN and the home network. Each of these entities could be classified as either 2G or 3G. Some of these entities are already classified as mixed cases, but from the security point of view it is useful to define a clear-cut division between 2G and 3G for each entity:

- The security module can either be a SIM card (2G case) or a UICC (3G case). It is important to note that a UICC may contain a SIM *application* in addition to a USIM *application*.
- The ME (Mobile Equipment) is classified as 2G if it supports exclusively the GSM RAN and interworks with either a SIM card or a SIM application in a UICC. Otherwise the ME is 3G: in which case it supports either UTRAN only or both GSM radio access and UMTS radio access. The 3G ME interworks with either a USIM application in a UICC or with a SIM.
- The division for RANs is clear: the GSM Base Station Subsystem (BSS) is used for 2G and UTRAN for 3G.
- The SN VLR/SGSN is classified as 2G if it supports exclusively GSM authentication and can be attached exclusively to a GSM BSS. Otherwise, the VLR/SGSN is 3G (i.e., it supports both the UMTS AKA and GSM AKA, and can be attached to UTRAN and/or a GSM BSS). Furthermore, a 3G SN supports conversion functions.
- The HLR/AuC is 2G if it supports exclusively authentication triplet generation for 2G subscriptions. A 3G HLR/AuC supports authentication quintet generation for 3G subscriptions and conversion functions to support GSM authentication. It may also support pure triplet generation for 2G subscriptions.

Altogether, we have $2^5 = 32$ different combinations of 2G/3G entities. If we also count the SIM application in the UICC as a third possible case for the security module, we have $3 \times 16 = 48$ cases (all these theoretical combinations are listed

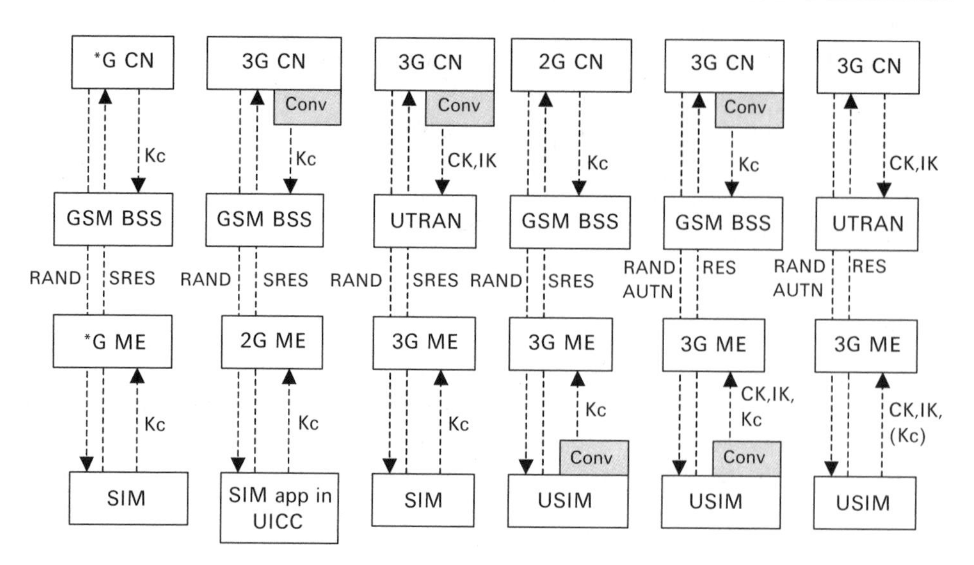

Figure 2.18 Interworking scenarios
CN = Core Network; CK = Cipher Key; IK = Integrity Key; GSM BSS = Global System for Mobile Base Station Subsystem; UTRAN = UMTS Terrestrial Radio Access Network; RAND = random number; SRES = user response in GSM; AUTN = authentication token; RES = user response; ME = Mobile Equipment; SIM = Subscriber Identity Module; UICC = Universal Integrated Circuit Card; USIM = Universal SIM

and analysed in [8]). In this book, we only highlight those scenarios that are essentially different from each other. To do this, we combine the CN entities in Figure 2.18 and say that the CN is 3G if either the SN or home network (or both) are 3G; otherwise, we say the CN is 2G. Six essentially different cases are depicted.

2.2.2 Cases with SIM

We have three essentially different cases where SIM is used as an access module.

2.2.2.1 SIM and GSM BSS

If SIM is used to access a GSM BSS, then we have a pure GSM case from the security point of view. It does not matter whether the ME is 3G or 2G and the same is true for the CN. As far as security features are concerned, we have 2G authentication and 2G encryption.

2.2.2.2 SIM application and GSM BSS

A slight variant of the previous case is when a UICC is used in 2G ME, when the RAN must be a GSM BSS. However, when SIM application is used in the UICC, exactly the same security features are carried out as in the previous case. In the CN, conversion functions must be available to produce authentication triplets. As far as security features are concerned, we have 2G authentication and 2G encryption.

2.2.2.3 SIM and UTRAN

In this case both the CN and the ME must be 3G, as they both support UTRAN. The GSM encryption key K_c is expanded to CK and IK by conversion functions both in the CN and in the ME. As far as security features are concerned, we have 2G authentication and both encryption and integrity protection are 3G but accessed by a 2G key.

2.2.3 Cases with USIM

We have again three essentially different cases when USIM is used as the security module. In all cases the ME must be 3G, since it must support USIM. For a similar reason, the home network must be 3G.

2.2.3.1 USIM and GSM BSS and 2G SN

Here the home network must produce authentication triplets with conversion functions because the SN can only support triplets. On the terminal side, USIM itself applies a conversion function to derive the GSM encryption key K_c. As far as security features are concerned, we have 2G authentication and 2G encryption.

2.2.3.2 USIM and GSM BSS and 3G SN

Once authentication vectors can be used, even if the RAN is only 2G, it is possible to run the UMTS AKA, as this protocol is transparent to the radio network. However, CK and IK cannot be used. Thus, a conversion function has to be used both in the USIM and in the CN to generate the GSM encryption key K_c. Note that CK and IK are transferred to the ME to support potential future handovers to UTRAN. On the security side, we now have 3G authentication but 2G encryption.

2.2.3.3 Pure 3G case

In this case all elements are 3G and the full set of UMTS security features are in use. Note that the converted GSM key K_c may be derived for potential future handovers to the GSM BSS. It would of course be technically possible to run GSM authentication in this case as well. Indeed, USIM has no way of knowing whether the ME is connected to UTRAN or GSM BSS. Therefore, the ME has to abort GSM authentication attempts in case it is connected to UTRAN and contains a USIM. It is only in this case that we have all the 3G security features: 3G authentication, 3G encryption and 3G integrity protection.

2.2.4 *Handovers from one system to another*

The concept of a handover is different for CS services than PS services. In PS it is much easier to send packets via different cells, but for a CS bit stream the transition from one cell to another has to be planned more carefully. This difference is also visible with inter-RAT (Radio Access Technology) handovers.

2.2.4.1 CS handovers from UTRAN to GSM BSS

The encryption algorithm must be changed during handover from UTRAN to the GSM BSS. The WCDMA algorithm UEA (UMTS Encryption Algorithm) is replaced by the GSM A5 algorithm and the UTRAN CK is replaced by a converted K_c. Information about supported/allowed GSM algorithms together with the key has to be transferred within the system infrastructure before the handover can take place. Of course, integrity protection is stopped at handover to GSM BSS.

2.2.4.2 CS handovers from GSM BSS to UTRAN

If the handover is done from GSM BSS to UTRAN, then the encryption algorithm is changed from A5 to UEA. Before the handover, GSM BSS requests UE to send information about its UTRAN security capabilities together with the associated parameters (e.g., CK, IK, START). This information is transferred within the system infrastructure to the target RNC before encryption and integrity protection can start on the UTRAN side.

2.2.4.3 Intersystem change for PS services

There are a couple of notable differences between intersystem handovers for CS services and corresponding intersystem changes to PS services. First, GPRS (General Packet Radio Service) encryption terminates in the CN and, therefore, transfer of keys is somewhat simpler. Second, there is a difference when the CN changes in addition to the radio network. If the UE moves to the area of a new MSC

(Mobile Switching Centre)/VLR, the old MSC/VLR still remains as the *anchor point* for the call. However, if the UE moves to the area of a new SGSN, then this new SGSN also becomes the anchor point for the connection.

2.3 Additional Security Features in Release 1999

There also exist other security features in the Release 1999 specification set: some are directly inherited from GSM as such and some are added for the first time in Release 1999. Let us now take a brief look at the individual features. Detailed information can be found in the relevant 3GPP specifications (see Table 1.2 for specifications under responsibility of the 3GPP security group SA3). Some relevant specifications (e.g., MExE (Mobile Execution Environment) or LCS (location services)) are not in this list, because security issues only partially cover them.

2.3.1 Ciphering indicator

There is a specific *ciphering indicator* in ME that is used to show the user whether encryption is applied or not, thus providing some visibility of the security mechanisms to the user. Note that although the use of ciphering is highly recommended it is still optional for the UMTS network. Details of the indicator are left to be implementation-specific and the best way to inform the user is very much dependent on the characteristics of the terminal itself (e.g., different display types may utilize different types of indicators).

In general, it is important that the security level is not dependent on whether the user is doing active checks. Nevertheless, for some specific actions, users may appreciate visibility of active security features.

2.3.2 Identification of the UE

In the GSM system, the ME can be identified by its International Mobile Equipment Identity (IMEI). This identity is not directly associated with the user because a SIM card may be moved from one terminal to another. There are, however, important features in the network that can only be based on the value of IMEI (e.g., it is possible to make *emergency calls* with a terminal without a SIM card). The only identification method in this case is to require the terminal to provide its IMEI, very useful also for tracing stolen phones.

This feature is carried over to the UMTS system as well. There are no mechanisms in either the GSM or UMTS that actually authenticate the provided IMEI. So, protection methods for IMEI have to be based solely on the terminal side: it must be made difficult for the terminal to be modified in such a way that it provides a wrong value for IMEI when requested by the network.

2.3.3 Security for Location Services (LCS)

The mobile network has to be able to trace users while they are on the move. Otherwise, it would not be possible to serve them. In addition to the needs of the network itself, there are clearly many services that can benefit from knowing the position of users (e.g., a hungry user may want to know which restaurants are closest to his or her current position).

Location information is clearly sensitive. People are not comfortable with the idea that they could be traced at any time. Security mechanisms have been defined to protect against leakage of location information to unauthorized parties. The *privacy profile* concept plays a central role here: the user must be in charge of who know about her or his whereabouts.

2.3.4 User-to-USIM authentication

This feature also carries over from the GSM system to the UMTS system and is based on a Personal Identification Number (PIN) known only to the user and the USIM. The user has to be able to give the PIN, which is 4–8 digits long, to the USIM before further access to the latter is granted. It must be admitted though that mobile phones are frequently stolen while they are in fully operational mode and therefore this feature does not act as a defence against theft, because authentication has already happened before the phone is stolen.

2.3.5 Security in the USIM application toolkit

Similarly to GSM, it is possible to build applications that are executed in the USIM by using a feature called the (U)SIM application toolkit, which grants the home operator the possibility, among others, to send messages directly to the USIM. The USIM application toolkit also specifies what kind of protection may be provided for this message transfer. Many details of protection mechanisms are implementation-specific.

2.3.6 Mobile Execution Environment (MExE)

The 3GPP has specified a framework for running applications in the ME (see [1]). Several different technologies are included in the specification (e.g., WAP (Wireless

Application Protocol) and Java). A great deal of the specification effort has been devoted to make the environment secure. In particular, security issues with downloaded applications have been addressed in [1]. Protection mechanisms are partially based on public key cryptography.

2.3.7 Lawful interception

In most countries, legislation and regulations set the requirement that authorities must have a way of accessing sensitive information (e.g., law enforcement has to be able to listen to the phone calls of suspected criminals or to find out where the suspects are (or were) at a certain moment). Such information is also used as evidence in court cases.

In the GSM, the lawful interception functionality was later added to an already existing complete system. Clearly, it is more effective to have standardized mechanisms in that kind of situation and this is why, in the UMTS, that the lawful interception features and the interfaces needed for them have been standardized as an integral part of the system.

When new elements and services are added to the 3GPP system, any lawful interception aspects are taken into account from the beginning. In this way it is possible to provide effective standardized solutions for this special purpose as well.

3

Security Features in Releases 4 and 5

3.1 Network Domain Security

The term "network domain security" in Third Generation Partnership Project (3GPP) specifications covers security of the communication between network elements. In particular, the User Equipment (UE) is not affected at all by network domain security. The two communicating network elements may be in the same network administered by a single mobile operator or may belong to two different networks. The latter case (i.e., internetwork communication) clearly requires standardized solutions, for otherwise each pair of operators that are roaming partners would need to agree separately on a common solution. The intranetwork case also benefits from standardization because network operators may have network elements manufactured by several different vendors.

In the past there have been no cryptographic security mechanisms available for internetwork communication. Security has been based on the fact that the global Signalling System No. 7 (SS7) network has only been accessible to a relatively small number of well-established institutions (e.g., network operators or large corporations). SS7 was standardized by the International Telecommunications Union (ITU) and still holds a major position in the fixed part of telecommunication networks. It has been very difficult for an attacker to insert or manipulate SS7 messages, but the situation is now changing for two reasons: first the number of different operators and service providers that need to communicate with each other is increasing; and second, there is a trend to replace SS7-based networks by Internet Protocol (IP)-based networks. The introduction of IP brings many benefits but it also means that a large number of hacking tools, some of which are available on the Internet, may become applicable to telecommunication networks (e.g., denial of service-type attacks).

For these reasons, the lack of cryptographic protection for internetwork communication may increasingly become a security risk for the current Global System for Mobile communications (GSM) system. In particular, session keys that are used to protect radio communications are sent in plaintext between operators. A major part of the 3GPP Release 1999 specifications was devoted to introduction of a completely new radio access technology while the Core Network (CN) part was an extension of the existing GSM specification set. This is the main reason why protection mechanisms for CN signalling were not introduced in Release 1999 but were in later releases, starting with Release 4.

The mobile-specific part of SS7 signalling is called the Mobile Application Part (MAP). In order to protect all communication in SS7 networks it is clearly not enough to protect the MAP protocol on its own. However, from the point of view of mobile communications, MAP is the essential part to be protected (e.g., the session keys that protect radio interface and other authentication data are carried in MAP). On the other hand, specifying a more general security protocol for SS7 would have been a major task and unlikely to be completed in the required time frame, at least not in time for Release 4. Mainly for these reasons, 3GPP has developed security mechanisms that are specific to MAP. The functional description of these mechanisms (stage 2) is given in TS 33.200 [12], while bit-level materialization (stage 3) is described in the MAP specification itself TS 29.002 [7]. The whole feature is called MAPsec and the first release in which it is included in 3GPP is Release 4. Note that the MAPsec protocol protects MAP messages at the application layer. We will give an overview of MAPsec in Section 3.1.1.

Many different security mechanisms have been standardized by the Internet Engineering Task Force (IETF) for IP-based networks. Hence, there is no need to specify anything from scratch in 3GPP. On the other hand, it is still important to agree on how IETF protocols are used to protect IP-based communication in 3GPP networks. Specification TS 33.210 [13] is devoted to this profiling task and is included in the 3GPP Release 5 specification set. The main tool from the IETF used in 3GPP is the IPsec protocol suite, see Request For Comments (RFCs) 2401–2412 [103]–[114].

Note that 3GPP also specifies how the MAP protocol can be run on top of IP. In this case, there are basically two alternative methods to protect MAP: MAPsec or IPsec. The latter has the advantage that protection also covers lower layer headers because protection is done in the IP layer.

3.1.1 MAPsec

The basic idea behind MAPsec can be described as follows. A plaintext MAP message is encrypted and the result is placed in a "container" inside another MAP message. At the same time a cryptographic checksum (i.e., a Message Authentication Code (MAC)) covering the original message is included in the new MAP message. To be able to use encryption and MACs, keys are needed. MAPsec has borrowed the notion of Security Association (SA) from IPsec. The SA not only contains cryptographic keys but also other relevant information (e.g., key lifetimes and algorithm identifiers). MAPsec SAs resemble IPsec SAs, but the two are not identical.

It is not specified in 3GPP Release 4 how SAs are to be exchanged between operators. In practice, this implies that the SAs have to be configured in the Network Elements (NEs) manually. Automatic key management for MAPsec has been designed by the 3GPP SA3 group, but at the time of writing it seemed probable that the detailed specification was not going to be included in the 3GPP Release 6 specification set. However, as the concept itself is fully stable in 3GPP, we present a high-level description of the planned functionality for automatic key management in this section.

The basic ingredient in MAPsec automatic key management is a new element called a Key Administration Centre (KAC). These KACs agree on SAs between themselves using the Internet Key Exchange (IKE) protocol. The KACs also distribute SAs to NEs. All elements in the same security domain (e.g., elements in one operator's network) share the same SAs and the policies necessary to handle these SAs and incoming messages. The sharing of SAs is unavoidable as only networks, and not individual NEs, can be addressed in MAP messages.

MAPsec (Figure 3.1) has three protection modes: no protection, integrity protection only and encryption with integrity protection. MAP messages in the last mode have the following structure: security header || f6(plaintext) || f7(security header || f6(plaintext)), where f6 is the Advanced Encryption Standard (AES) algorithm in counter mode and f7 is AES in Cipher Block Chaining (CBC) MAC mode. The security header contains information needed to be able to process the message at the receiving end (e.g., security parameter index, sending the NE identifier and the time-variant parameter).

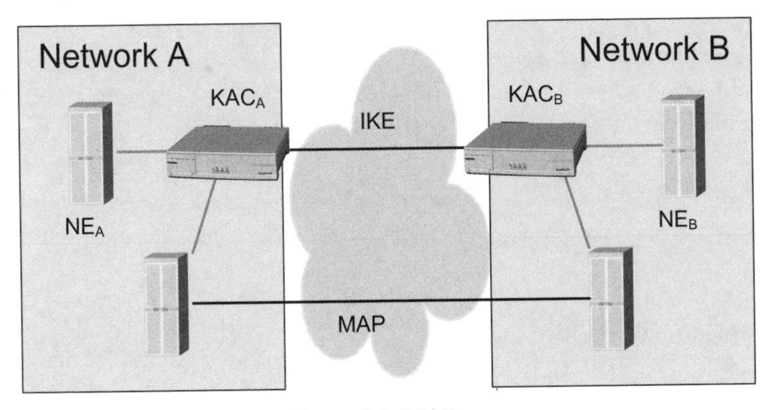

Figure 3.1 MAPsec
MAP = Mobile Applicaton Part; KAC = Key Administration Centre; IKE = Internet Key Exchange; NE = Network Element

3.1.1.1 General principles of MAPsec

MAPsec is introduced to add protection to the existing global network of MAP-capable NEs. This implies that there has to be an unavoidable transition period during which an increasing number of NEs support MAPsec, while others do not. While this is the case, MAPsec-capable elements must communicate with elements that do not support MAPsec, giving a clear advantage for an active attacker, as she will have the chance to masquerade as an NE without MAPsec. Protection against active attackers will continue to be considerably limited until all networks support MAPsec. On the other hand, protection against passive attacks (e.g., trying to catch authentication vectors by eavesdropping MAP communications) increases each time a new network begins to use MAPsec. If the communication is encrypted then a passive attack does not work.

Unfortunately, protection against passive attacks does not increase in linear proportion to the increased number of MAPsec-capable NEs. Let us look at a simplified scenario in which there are n networks of equal size that communicate with each other in roughly equal volume. If half of the networks are MAPsec-capable, then we have roughly $n^2/4$ protected communication "lines" out of a total of roughly n^2.

These remarks are not only applicable to MAPsec, a similar situation appears in any existing communication network where protection is introduced gradually. Special methods should be used throughout the transition period to guarantee that unprotected communication is at least monitored with care and, where possible, all actions caused by unprotected messages are double-checked before execution.

A further complication regarding this transition period comes about because there is another transition toward IP-based transport that is running in parallel. Hence, the use of IPSec protection for MAP traffic could replace the use of MAPsec; there is only a small advantage in using both protection mechanisms at the same time. This puts another requirement on MAPsec policy management. The MAP layer must be aware at the same time whether IPsec protection is in use and whether IPsec policies are acceptable from the MAPsec point of view. In principle, there could be a situation where traffic is protected by IPsec but still contains unacceptable information elements at the MAP layer. These situations can only be avoided by some kind of interplay between the MAP and IP layers inside the communication nodes.

3.1.1.2 Structure of MAPsec messages

There are three protection modes in MAPsec: protection mode 0 gives no protection; protection mode 1 provides integrity protection (and thus message authentication);

and protection mode 2 provides both integrity protection and encryption. Note that the use of encryption without integrity protection is not possible.

In all three modes, a MAPsec-protected message consists of a *security header* and a *protected payload.* The security header is always sent in cleartext, simply because it is important to process MAPsec correctly at the receiving end. The protected payload contains essentially the original MAP message payload in protected form. In mode 1, a MAC is calculated over the security header and original payload and the result is appended to the MAPsec message. In mode 2, MAC is calculated over the security header and the encrypted payload.

The security header consists of the following data elements (in mode 0, only the first two are used):

$$\textit{Security header} = \textit{SPI} \parallel \textit{Original component ID} \parallel \textit{TVP} \parallel \textit{NE-Id} \parallel \textit{Prop}$$

where *SPI* is a Security Parameter Index that, together with destination PLMN (Public Land Mobile Network) identity, points to a unique MAPsec SA; *Original component ID* refers to the type of original MAP message (needed in cleartext to be able to process MAP correctly); *TVP* is a Time Variant Parameter that is needed to provide protection against replay attacks; *NE-ID* identifies the sending network element; and *Prop* is a proprietary field intended for local use when creating the Initialization Vector (IV) needed in encryption and MAC algorithms.

3.1.1.3 MAPsec algorithms

MAPsec allows the use of several encryption algorithms, which is necessary for future-proofing purposes. However, only one algorithm was specified before Release 5, in addition to the NULL algorithm (i.e., the identity function). This is called MEA-1 (MAPsec Encryption Algorithm) and is equivalent to AES in counter mode (see [13], [90] and [92] for further details).

For integrity protection, an algorithm defined in [61] is used in addition to the NULL algorithm (appended to an empty MAC).

The initialization vector consists of *TVP*, *NE-Id* and *Prop* padded with 0 bits.

3.1.1.4 Protection profiles

In MAPsec only some MAP operations are protected (i.e., the most critical ones like *authentication data transfer* or *reset*). This is for performance reasons. Furthermore, different components in a MAP operation may have different protection modes. These properties of MAPsec have led to specification of a notion called a *protection profile*. Each protection profile defines both the extent of protection and protection

Table 3.1 MAPsec protection profiles

Protection profile name	Protection group				
	PG(0) No protection	PG(1) Reset	PG(2) AuthInfo except handover situations	PG(3) AuthInfo in handover situation	PG(4) Nonlocation-dependent HLR data
Profile A	✓				
Profile B		✓	✓		
Profile C		✓	✓	✓	
Profile D		✓	✓	✓	✓
Profile E		✓	✓		✓

PG = Protection Group; HLR = Home Location Register; AuthInfo = Authentication Information

Table 3.2 MAPsec protection group 2

InfoRetrievalContext-v3/Send Authentication Info
InfoRetrievalContext-v2/Send Authentication Info
InfoRetrievalContext-v1/Send Parameters
InterVlrInfoRetrievalContext-v3/Send Identification
InterVlrInfoRetrievalContext-v2/Send Identification

modes for each protected MAP component. An auxiliary concept of a *Protection Group* (PG) is also used. In Table 3.1 profiles defined in 3GPP Release 5 are given.

As an example of a protection group we present group 2, which extends to application context/operation pairs as shown in Table 3.2.

The protection mode for the *invoke* component is mode 1 in all these cases, while the protection mode for the *result* component is 2 and for the *error* component it is 0. Note that the result component contains the actual authentication data. A *protection profile revision identifier* is introduced for increased coverage of the MAPsec protection at a later time. The profiles defined in Release 5 constitute revision number 0.

3.1.1.5 Security associations

MAPsec also uses SAs, which were originally created for IPsec. In addition to the keys needed for cryptographic operations, an SA contains other relevant information so that the key can be used in the correct manner.

An SA contains the following data elements:

- *destination PLMN ID*;
- *SPI*;
- *sending PLMN ID*;
- *encryption key*;
- *encryption algorithm*;
- *integrity key*;
- *integrity algorithm*;
- *protection profile ID*;
- *protection profile revision ID*;
- *soft expiry time*;
- *hard expiry time*.

The first two elements have already been explained in the context of the security header and together they point to a unique SA. This is guaranteed by the rule that an SPI is always chosen at the receiving end. The other attributes (except the last two) have already been explained in previous sections. Both expiry times are expressed in *Co-ordinated Universal Time* (UTC), the worldwide time system. Of course, the accuracy of local time determines how well all the actions imposed by the expiry times can be executed. After a soft expiry time has been reached, the SA should not be used at the sending end unless it is the only valid SA available. The receiving end may freely apply the SA, even after the soft expiry time. After reaching the hard expiry time, the SA cannot be used at all. The distinction between the two lifetimes is needed so that the SAs in use can change smoothly while communication continues without interruptions. A new SA has to be negotiated well ahead of the soft expiry time so that both parties can begin to use it immediately after the soft expiry time. It is important to take care that the sending end does not have the new SA available before it is available at the receiving end; otherwise, it is possible that at the very moment of soft expiry time the sending end starts to use an SA that is not yet available at the receiving end.

It is difficult to specify exactly how much in advance of soft expiry time that negotiation of a new SA must begin. However, because we have two expiry times there is some leeway that enables the continuous function of MAPsec in case the negotiation is started too late.

All SAs are stored in an SA Database (SAD) and all MAPsec NEs must have access to it.

3.1.1.6 Automatic key management

It is possible to operate MAPsec without any standardized mechanisms to support key or SA management. However, this easily leads to use of SAs with very long lifetimes (e.g., one year). The keys are used across the entire network, and their compromise can have serious consequences. The specification for MAPsec automatic key management is not included in Release 6, but the details of the mechanisms described in this section will probably be included in a later release specification of 3GPP. The principles underlying the mechanisms were agreed at the 3GPP SA3.

In Release 4 MAPsec contains the specification of the Zf-interface (Figure 3.2). A new MAP application called "secure transport" is introduced for this purpose, where automatic key management consists of specification of Zd- and Ze-interfaces. The former is an internetwork interface and must be standardized. The latter is an intranetwork interface that is also standardized because many mobile operators run a multivendor network.

A KAC is used to negotiate SAs on behalf of NEs. IKE [111] and MAPsec Domain of Interpretation (DoI) [35] are planned to be used for this. All SAs are valid on a PLMN-to-PLMN basis. The main reason for this sharing of SAs is that a PLMN can only address another PLMN and not any of its individual NEs.

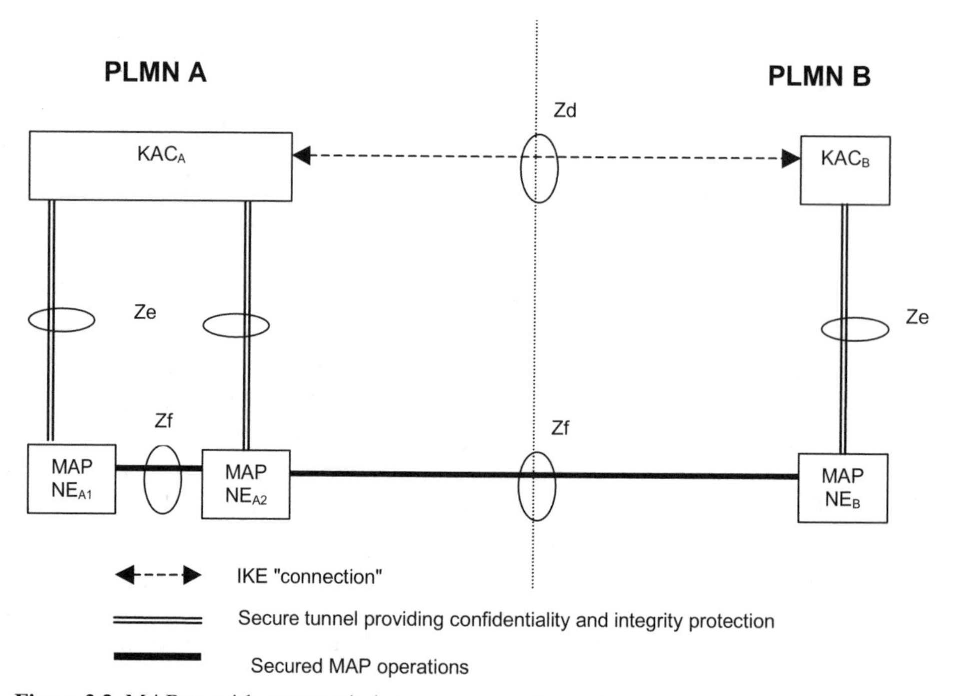

Figure 3.2 MAPsec with automatic key management
MAP = Mobile Application Part; PLMN = Public Land Mobile Network; KAC = Key Administration Centre; NE = Network Element; IKE = Internet Key Exchange

KAC also maintains an SAD and a Security Policy Database (SPD) (similar databases are also maintained locally in NEs). SAs and policies are distributed from the KAC to NEs over the Ze-interface. When an NE needs to address another NE, potentially in other network, it begins by looking up the local SPD. If policy information about communication already exists, then the NE acts according to the policy; otherwise, policy information has to be fetched from the KAC in advance.

Let us assume that secure transport is mandated in the SPD and that the NE looks at the SAD for a valid SA. If such an SA exists, then the NE may start using MAPsec over the Zf-interface; otherwise, an SA is requested from the KAC. If a valid SA exists in the SAD of the KAC, then that is transferred to the NE. If not, then the KAC contacts its peer in the other PLMN and begins the SA negotiation process over the Zd-interface. Of course, this last phase can only occur in the inter-PLMN case.

If SAs expire in the KAC SAD, then the KAC does not have to wait until some NE requests a new SA; instead, the KAC may act proactively. Entries in the SPD have to be filled in internally in each PLMN. Naturally, the contents of the SPD have to reflect the capabilities of peer PLMNs. An important item in the SPD is an indicator called "fallback to unprotected mode", which shows whether unprotected communication is allowed or not. In general, the SPD should show which protection mode is assumed for which MAP operation toward each peer PLMN. As already discussed, it is important that policies toward other networks are as uniform as possible; otherwise, active attackers would try to find those PLMNs where unprotected communication is allowed.

There may be several KACs in the same PLMN. This is useful for load balancing and to avoid a single point of failure. Clearly, the KAC is a sensitive element in the network and a natural target for attacks, so both it and its databases have to be physically secured.

3.1.2 IPsec

We will now give a brief overview of IPsec as specified in RFCs 2401–2412. IPSec mechanisms are used in the 3GPP security architecture both for network domain security of IP-based networks and for IMS (IP Multimedia CN Subsystem) access security.

IPsec is standardized by the IETF and consists of a dozen RFCs. It is a mandatory part of IPv6. In IPv4, IPsec can be used as an optional "add-on" mechanism to provide security in the IP layer. The main IPsec components are the following:

- Authentication Header (AH);

- Encapsulation Security Payload (ESP);
- IKE.

3.1.2.1 How IPSec works

The purpose of IPsec is to protect IP packets: this is done by means of the ESP and/or the AH. In short, ESP provides both confidentiality and integrity protection while AH only provides the latter. There are more fine-grained differences between the two, but clearly ESP and AH are largely overlapping mechanisms. One of the reasons for allowing this kind of redundancy in IPsec standards is export control: there have been severe restrictions on the export of confidentiality protection mechanisms in most countries, while integrity protection mechanisms have typically been free from restrictions. Currently such export restrictions have been eased and, as a consequence, the importance of AH compared with ESP is decreasing.

Both ESP and AH need keys. More generally, though, SAs are essential to IPsec. In addition to the encryption and authentication keys, SA contains information about the used algorithm, lifetime of the keys and the SA itself. It also contains a sequence number to protect against replay attacks, etc.

SAs must be negotiated before ESP or AH can be used, and one SA is needed for each direction of communication. This is done in a secure way by means of the IKE protocol. There are several IKE modes but the basic idea is the following: the communicating parties are able to generate "working keys" and SAs, which are used to protect subsequent communication. IKE is based on the ingenious idea of public key cryptography where secret keys for secure communication can be exchanged over an insecure channel. However, authentication of the parties who run IKE cannot be done without some long-term keys. These are typically based on either manual exchange of a (pre-) shared secret or, alternatively, on the Public Key Infrastructure (PKI) and *certificates*. Both solutions are clearly nontrivial to deploy: the former requires lots of configuration effort while the latter implies dedicated infrastructure elements with special functionality.

The negotiation of SAs by IKE is independent of the purpose for which these SAs are used. This is why IKE can also be used for negotiation of keys and SAs in MAPsec.

3.1.2.2 IPsec security associations

The most important parameters in an SA are [103]:

- the authentication algorithm (in AH/ESP);

- the encryption algorithm (in ESP);
- the encryption and authentication keys;
- lifetime of encryption keys;
- the lifetime of the SA itself;
- replay protection sequence number.

SAs are maintained in a specific SAD and they may be uniquely identified by the parameters SPI, destination IP address and IPsec protocol (AH/ESP).

3.1.2.3 ESP structure

There is a fair amount of overlap between AH and ESP, which is the reason we only explain ESP here in detail. As mentioned earlier, ESP provides both ciphering and integrity protection while AH is only about integrity protection, but there are also some other differences: most importantly, AH authenticates the IP header of the packet.

Let us now describe ESP in more detail. There are two ESP modes: *transport* mode and *tunnel* mode. Transport mode functions are basically as follows. Everything in an IP packet except the IP header is encrypted. Then a new ESP header is added between the IP header and the encrypted part, which contains such information as the SPI. Also, the encryption typically adds some bits into the end of the packet. Finally, an MAC is calculated over everything except the IP header and is appended to the end of the packet. At the receiving end, the integrity is checked first. This is done by removing the IP header from the beginning of the packet and the MAC from the end of the packet, then running the MAC function (using the algorithm found based on the information in the ESP header) over the rest of the packet and comparing the result with the MAC in the packet. If the outcome of the integrity check is positive, then the ESP header is removed and the rest is decrypted (again based on information in the ESP header) (see Figure 3.3 for illustration).

Tunnel mode differs from transport mode in the following way. A new IP header is added to the beginning of the packet. Then the same operations as in transport mode are done for the new packet. This means that the IP header of the original packet is also protected (as illustrated in Figure 3.3).

Transport mode is the basic use case of ESP between two end points. However, when applied in 3GPP networks, communicating NEs need to:

- know the IP address of each other;
- implement the IPsec functionality.

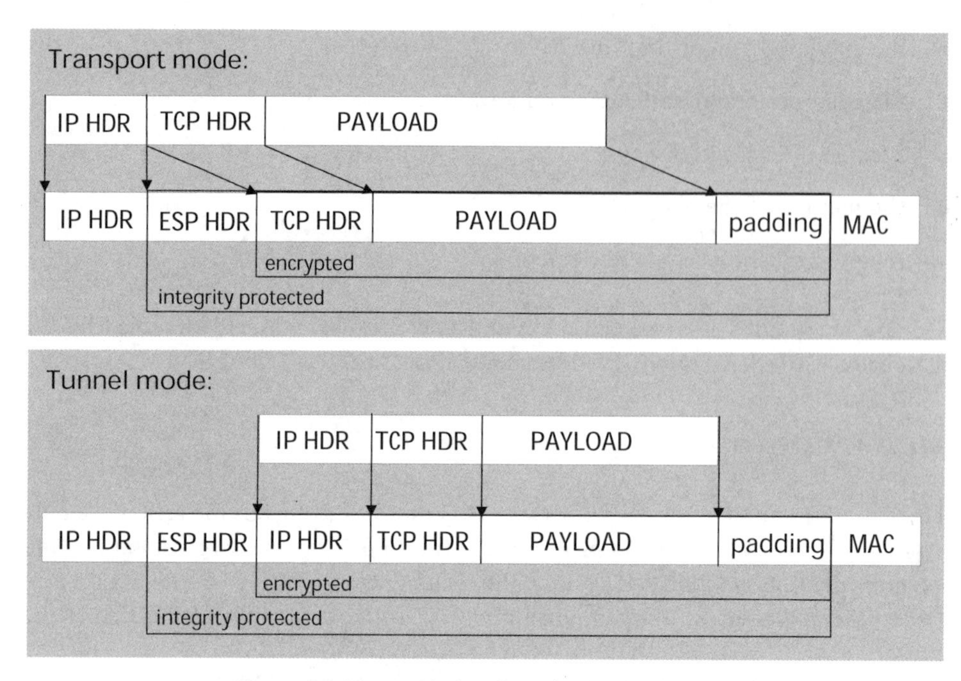

Figure 3.3 Encapsulation Security Payload (ESP)
IP = Internet Protocol; TCP = Transmission Control Protocol; MAC = Message Authentication Code

The typical use case of tunnel mode is related to the concept of a Virtual Private Network (VPN). IPsec is used between two middle nodes (security gateways or VPN gateways), and end-to-end protection is provided implicitly because the whole end-to-end packet is inside the payload of the packet that is protected between the gateways. Clearly, in this scenario each element has to trust the corresponding gateway. In addition, the leg from the end point element to the gateway has to be protected by other means (e.g., implemented in a physically protected environment). The preferred protection method for UMTS (Universal Mobile Telecommunications System) CN control messages is to use ESP in tunnel mode between security gateways.

3.1.3 IPsec-based mechanisms in UMTS

A basic part of the IPsec-based security in 3GPP systems is the *security gateway*. All control plane IP communications toward external networks should go via security gateways (Figure 3.4). These gateways use the IKE protocol [111] to exchange IPsec SAs between themselves. An important conceptual distinction between a security

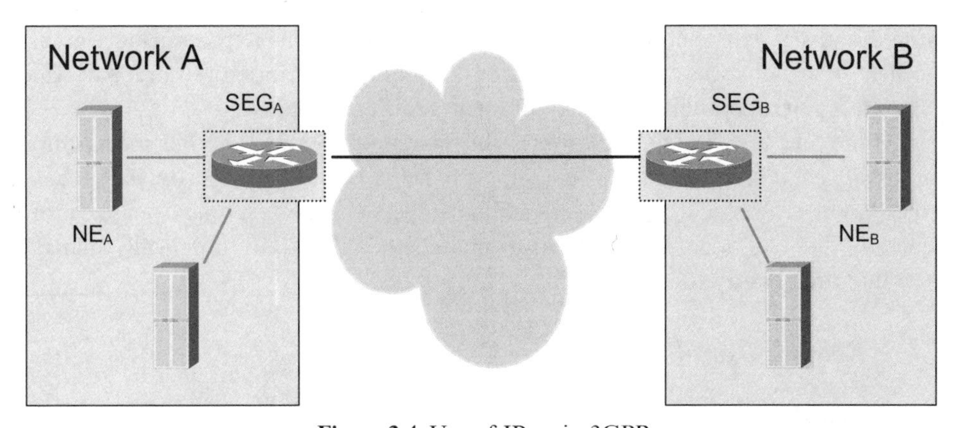

Figure 3.4 Use of IPsec in 3GPP
IP = Internet Protocol; 3GPP = Third Generation Partnership Project; SEG = Security Gateway; NE = Network Element

gateway and a MAPsec KAC is that the former also uses negotiated SAs itself, while the latter distributes them to other elements, which send and receive the actual MAPsec messages. In Release 5 the IKE is based on preshared secrets, but in future releases support of a PKI for key management may be added.

A security gateway contains both an SAD and an SPD, which indicate how and when SAs are used or have to be used. Naturally, the security gateway has to be physically secured and, typically, would be combined with firewall functionality.

One obstacle to reaching full interoperability of IPsec is the great number of options in the specifications. 3GPP has tried to solve this problem by cutting down the number of options, which can be summarized as follows:

- ESP [108] is only used for protection of packets, while AH [104] is not used at all;
- ESP is always used in tunnel mode;
- triple DES (Data Encryption Standard) is chosen as the encryption algorithm (the newer algorithm AES published by NIST was considered for a long time in the specification work, but no IETF RFC to support use of AES in IPsec appeared within the Release 5 time frame);
- the IV is always generated at random;
- HMAC-SHA-1 (keyed-Hashing MAC with Secure Hash Algorithm #1) is the mandatory message authentication algorithm;
- IKE is used for key exchange and within IKE the chosen option is to use the main mode in phase 1 with preshared secrets.

Note, however, that operators might configure more options (e.g., transport mode) in their own networks. The specification [13] only describes the core part that guarantees interoperability between different security domains.

At the time of writing, 3GPP was in the process of creating another specification for an *authentication framework* that supports network domain security. The crucial concept here is a specially tailored PKI that can be used between operators (or, more generally, between security domains) to make it possible to automatically manage keys that are already used for IKE.

3.1.4 Role of firewalls

Access control is one of the basic techniques in network security systems. The idea of access control is to restrict traffic into/out of a protected area in the network. A typical example is protection of traffic between public Internet and mobile operator networks. The goal of access control is to let only authorized traffic pass through; all other traffic is rejected. An obvious trade-off between availability and usability, on one side, and security, on the other side, exists here: maximal protection is achieved if all traffic is simply rejected, but then we get zero-level availability. On the other hand, by allowing all traffic to reach our network we provide maximal availability for all users, including hostile intruders and attackers. However, the consequence of malicious actions may be that the whole system falls down. Similar trade-offs are typical for many other security issues as well.

Access control can be applied to many different levels. On the network layer it is possible to check traffic packet by packet (e.g., based on IP addresses) and only let packets from allowed addresses to go through. Going to higher layers, it is possible to restrict traffic based on the protocol headers (e.g., only certain protocols are allowed through the access control point). These are examples of a technique called *packet filtering*.

In the application layer, access control is typically carried out by examining a number of user IDs. Only authorized users are allowed to go through and use resources behind the access control point. This is usually materialized in the form of Access Control Lists (ACLs).

The firewall is a computer (machine/network element) that executes access control at the border of the protected network. Methods at different layers are usually applied, filtering functions in lower layers first and then proxy-type functionality in higher layers, where protocol messages are analysed at the firewall and forwarded if they pass access control.

The rules that define which traffic gets the green light are of utmost importance for a firewall to be effective, as illustrated by the two extreme examples above (rejecting or allowing all traffic). Indeed, security policies play a major role in the context of firewalls and access control. The main weakness of firewall-type approaches is that once the traffic passes the firewall it can go freely into any other element behind the firewall. Of course, the philosophy is based on optimization of cost: we only need one heavily protected gateway that guards the whole network. But it is very difficult to measure the would-be effect of each incoming packet to the network. To be able to estimate this fully, the firewall should ideally be able to emulate the whole network behaviour while not degrading the efficiency of network use by legitimate users.

A firewall resides at the border of a network. To define exactly what a border is and to determine whether a firewall is inside or outside the network it protects, a Demilitarized Zone (DMZ) is introduced, an area that neither belongs to the "inside" network nor to the "outside" network. A firewall is a typical entity that resides in the DMZ: anybody from "outside" the network can access it directly and it can access the "inside" of the network, but there are no direct connections between "outside" and "inside". Other entities that reside in the DMZ include FTP (File Transfer Protocol) or HTTP (Hyper Text Transfer Protocol) servers that are open to public access.

3.2 IMS Security

A major part of 3GPP Release 5 is devoted to the specification of the IP Multimedia CN Subsystem (IMS). It is a complete application layer system that is built on top of the UMTS PS (Packet Switched) domain, but is designed in such a way that it is independent of the underlying access technology. In addition to UTRAN (UMTS Terrestrial Radio Access Network) and GERAN (GSM/EDGE Radio Access Network) access, it is planned for future releases to support other kinds of access technologies (e.g., WLAN [Wireless Local Area Network] access to IMS is possible).

The crucial protocol in IMS is the Session Initiation Protocol (SIP), which is specified in IETF [120]. SIP manages IP-based sessions by setting up, modifying and terminating multimedia sessions. User plane traffic is separated from control plane traffic in IMS. In Release 5's version of IMS, the aim is just to provide basic functionality (i.e., a platform on top of which various IP-based services can be developed). In Release 6, some IMS services are going to be standardized (e.g., presence, push, instant messaging and chat). Later, conversational real-time services like video-conferencing will be the target. In addition to SIP, IMS utilizes the Session Description Protocol (SDP) to negotiate the parameters used in sessions (see [102] and [121]).

3.2.1 Basics of SIP

In a nutshell, SIP works as follows. A user (or, more technically, a User Agent [UA]) sends INVITE messages to other users in order to initiate sessions where multimedia

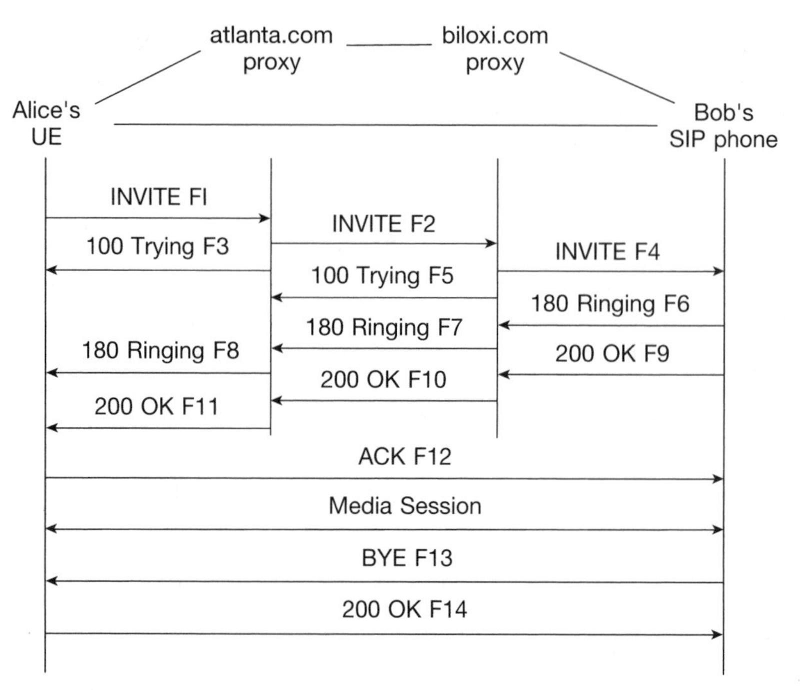

Figure 3.5 SIP session set-up example using a SIP trapezoid
SIP = Session Initiation Protocol

data are exchanged. SIP proxies help users in this task. UAs also send REGISTER messages to SIP servers that are called *registrars*. This registration of a UA helps other users to find it. The invited UA sends an OK message back to the inviting UA if it decides to accept the invitation to a session. In the payload of SIP messages, properties of the session media are described using SDP. These properties include name and time of the session, type and format of the media streams, addresses and ports for receipt of the media, etc. Users may terminate the session by sending BYE messages.

Figure 3.5 shows how RFC 3261 illustrates a basic SIP session. In addition to messages described in the paragraph above, there are other response messages (e.g., indicating ringing) that inform participants about the state of the session set-up.

SIP is based on a *request–response* model, similarly to HTTP [116]. Every SIP message is either a request sent from a *client* to a *server* or a response from a server to a client. Note that UAs contain both client (UAC) and server (UAS) functionalities.

There are six basic request types, called *methods* in SIP: REGISTER for registering the contact information of a user; INVITE, ACK and CANCEL for setting up sessions; BYE for terminating sessions; OPTIONS for finding out the capabilities of the server. Several extension methods have also been specified (e.g., INFO for carrying mid-session signalling information [119] and MESSAGE for instant messaging [124]).

Responses contain a *Status-Code* made up of a three-digit integer. The first digit indicates the type of response in the following way:

- 1xx is a *provisional* response that takes place while processing of the request continues;
- 2xx indicates *successful* reception of the request;
- 3xx indicates *redirection* (i.e., further action is needed in order to complete the request);
- 4xx indicates *client error*—either the request is somehow badly constructed or sent to the wrong server;
- 5xx indicates *server error* (i.e., the request was valid but the server cannot fulfil it);
- 6xx is a *global failure*—the request cannot be fulfilled by any server.

A *transaction* consists of a request sent by a client and all responses to the request sent back by the server. The SIP *transaction layer* takes care of (application layer) retransmissions of requests and responses, matching of responses to correct requests and (application layer) timeouts.

The INVITE request establishes a *dialog*: a peer-to-peer SIP relationship that persists for some time and comprises several transactions. A dialog also facilitates sequencing and proper routing of SIP messages between the UAs.

In addition to method names, SIP messages consist of *header fields* and *message bodies*. Some examples of header types are listed below:

- *Via* contains the address to which responses should be routed;
- *To* specifies the logical name of the recipient of the request;
- *From* indicates the initiator of the request;
- *Call-ID* uniquely identifies a session of a particular user;
- *Contact* points to a URI (Uniform Resource Identifier) and its meaning depends on the type of request;
- *Content-Type* indicates what kinds of media are carried in the message body;
- *Authentication-Info* is utilized for mutual authentication by the HTTP Digest mechanism [117];
- *Authorization* contains the credentials of the UA needed for authentication (e.g., a cryptographic response to a challenge);

- *Priority* indicates the urgency of the request;
- *Record-Route* is inserted by a proxy to force future requests to be routed through the same proxy;
- *Subject* indicates the nature of the call.

3.2.2 IMS architecture

The IMS architecture is defined in [3]. The central elements of IMS are a number of SIP servers and proxies, called Call Session Control Functions (CSCFs). The main purposes of these are listed in the following:

- The proxy CSCF (P-CSCF) is the first contact point in IMS. It carries out the following functions:
  - it forwards SIP registration requests received from the UE to an I-CSCF;
  - it forwards SIP messages received from the UE to the SIP server (e.g., S-CSCFs) whose name the P-CSCF has received as a result of the registration procedure;
  - it forwards SIP requests or responses to the UE;
  - it should perform SIP message compression/decompression.
- The interrogating CSCF (I-CSCF) is the contact point within a subscriber's home network. Its functions are:
  - as part of the registration procedure, to assign an S-CSCF to a user;
  - to route an SIP request received from another network toward the S-CSCF;
  - to forward SIP requests or responses to the S-CSCF.
- The serving CSCF (S-CSCF) undertakes session control services for the UE by maintaining a session state. The functions it carries out during a session are listed as follows:
  - it accepts registration requests and informs the Home Subscriber Server (HSS);
  - session control for the registered user's sessions;
  - interaction with service platforms for the support of various services;

- it provides end points with service event-related information (e.g., notification of tones);
- on behalf of both originating and terminating subscribers, it forwards SIP requests and responses to I-CSCF or P-CSCF (or to other elements in case one of the communicating parties is outside the IMS).

In addition, all CSCFs may play a role in charging.

3.2.3 Architecture for securing access to the IMS

Security features must clearly be part of the basic platform and they are already specified in Release 5. All security features of Release 5's IMS are described in Figure 3.6. These are specified in 3GPP TS 33.203 [136].

When a UA wants to access the IMS, it first creates a PDP (Packet Data Protocol) context with the PS domain. In this process, UMTS Release 1999 security features are utilized: mutual authentication between the UE and the PS domain, integrity protection and encryption between the UE and the RNC (Radio Network Controller). Through the GGSN (Gateway GPRS [General Packet Radio Service] Support Node), the UA is able to contact IMS nodes using SIP signalling. The first contact in IMS is P-CSCF. Through it the UA is able to register itself to the home IMS. At the same time the UA and the home IMS authenticate each other, based on a permanent, shared, master secret. They also agree on temporary keys to be used for further protection of SIP messages.

The SIP traffic between the visited IMS and the home IMS is protected by network domain security mechanisms. The SAs used for this purpose are not specific to the UA in question.

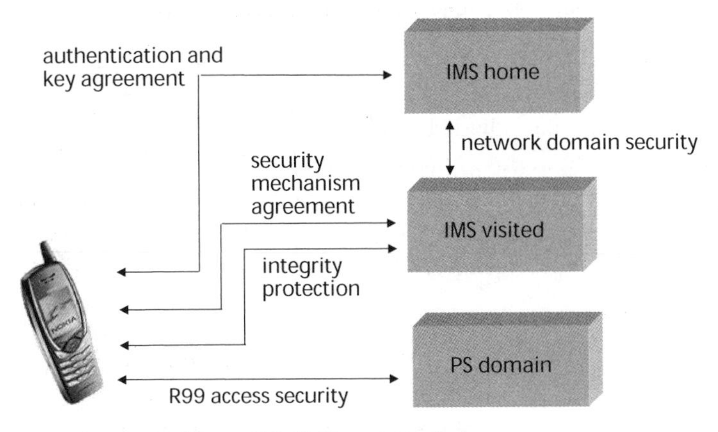

Figure 3.6 IMS security features
IMS = IP Multimedia CN Subsystem; IP = Internet Protocol; CN = Core Network; PS = Packet Switched

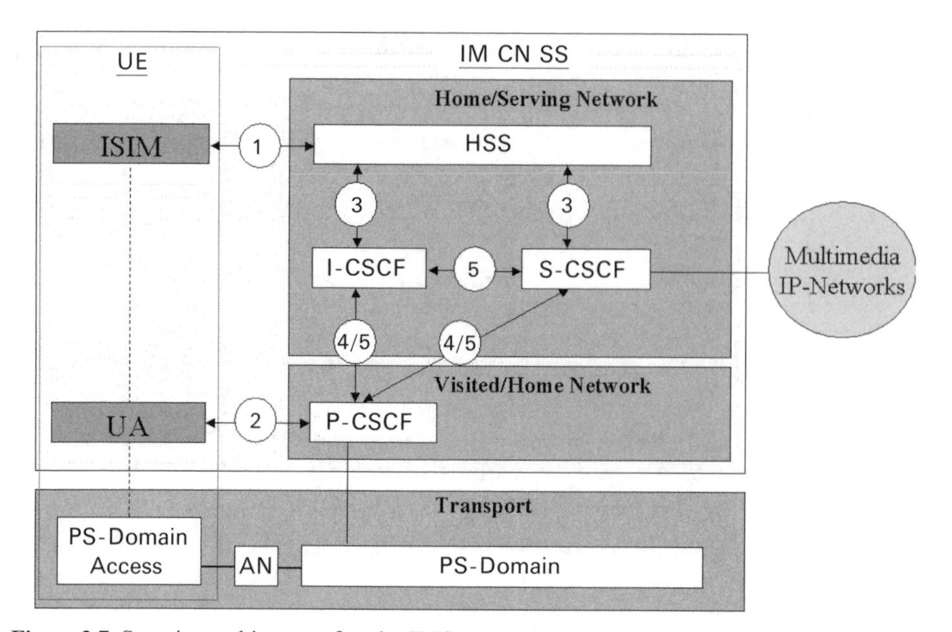

Figure 3.7 Security architecture for the IMS
IMS = IP Multimedia CN Subsystem; IP = Internet Protocol; CN = Core Network; UE = User Equipment; IM CN SS = IP Multimedia Core Network Subsystem; ISIM = IMS Subscriber Identity Module; HSS = Home Subscriber Server; CSCF = Call Session Control Function; UA = User Agent; PS = Packet Switched

Next, the UA and the P-CSCF negotiate *in a secure manner* all the parameters of the security mechanisms to be used to protect further SIP signalling (e.g., cryptographic algorithms are negotiated). Finally, integrity protection of all first-hop SIP signalling between the UA and P-CSCF is started, based on the temporary keys agreed during the authentication phase. The various security contexts needed for the IMS security architecture are shown in Figure 3.7.

The security architecture relies on three main components:

1. A permanent security context between the UE and the HSS, achieved by using a security module called the IMS Subscriber Identity Module (ISIM) that dwells in a tamper-resistant environment on the user's side and, on the network side, the HSS has a secure database. Both ISIM and HSS contain the IMS subscriber's identity and a corresponding master key based on which the Authentication and Key Agreement (AKA) procedure can be carried out.

2. A temporary security context exists between the terminal equipment on the user's side and a P-CSCF on the network side, consisting of IPsec ESP SAs and a link between these SAs and IMS subscribers. The security context is used to authenticate each SIP message individually in the first hop (from the UE to the P-CSCF).

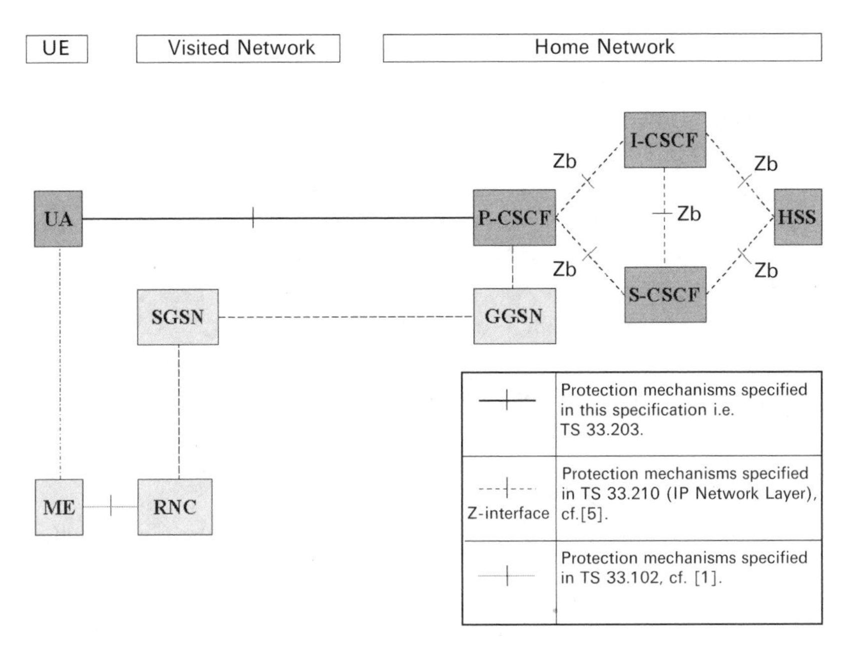

Figure 3.8 Network domain security in IMS
IMS = IP Multimedia CN Subsystem; UE = User Equipment; UA = User Agent; I-CSCF = Interrogating Call Session Control Function; P-CSCF = Proxy CSCF; S-CSCF = Serving CSCF; HSS = Home Subscriber Server; SGSN = Serving GPRS Support Node; GPRS = General Packet Radio Service; GGSN = Gateway GPRS Support Node; ME = Mobile Equipment; RNC = Radio Network Controller

3. All control plane (both SIP and other protocols) traffic between different network nodes is protected using mechanisms described in Section 3.1 (Network Domain Security (NDS)) (see Figure 3.8).

Note that SIP RFC 3261 mandates the use of Transport Layer Security (TLS) (see [100]), for all SIP servers and registrars. In IMS, however, security cannot solely be based on TLS, as TLS necessitates Transmission Control Protocol (TCP), which cannot always be used in the wireless environment. Instead, UDP is often used as the transport protocol and, hence, is the main reason TLS does not have a central role in the IMS security architecture.

3.2.4 Principles for IMS access security

In this section we give an overview about how components of the IMS security architecture are used in SIP (Figure 3.9).

There are two SIP procedures that play a central role both in SIP itself and in IMS security: REGISTER for registration and INVITE for session establishment. We describe the security solution for IMS around these two basic functions.

Use of IMS is based on a subscription. The user makes an agreement with the IMS operator and has an IMS Private Identity (IMPI) stored in both the ISIM and the HSS. There is also a cryptographic, 128-bit master key stored in association of the IMPI. It is not intended that IMPI be used to address the user; instead, there exists at least one IMS Public Identity (IMPU) that is tied to IMPI. There may be different service profiles inside a single subscription that result in different IMPUs being tied to the same IMPI (e.g., there may be different profiles for business and leisure time use). Technically, IMPI has the form of a Network Access Identifier (NAI), as defined in [115], while IMPU has the form of a SIP URI [120] or a *tel* URL (Uniform Resource Locator) [118].

Before a subscriber can begin to use services provided by the IMS, she must have an active registration, which can be obtained by sending a REGISTER request message to a P-CSCF. This message contains both the private address IMPI to be authenticated and at least one public identity IMPU to be registered. The problem of finding the address of an appropriate P-CSCF is solved by signalling in the underlying UMTS PS domain (in Release 5).

The P-CSCF forwards the REGISTER request to the I-CSCF, which in turn contacts the HSS to allocate a suitable S-CSCF to the user. All this communication, as well as all subsequent communication between NEs, is protected by NDS methods using SAs that are not specific to the subscriber.

After the S-CSCF is selected, the REGISTER message is forwarded to it. Next, S-CSCF fetches Authentication Vectors (AVs) from the HSS. Note that the IMS AKA is here based on the same mechanism that is used for a similar purpose in CS (Circuit Switched) and PS domains. Furthermore, it is also possible to reuse the same

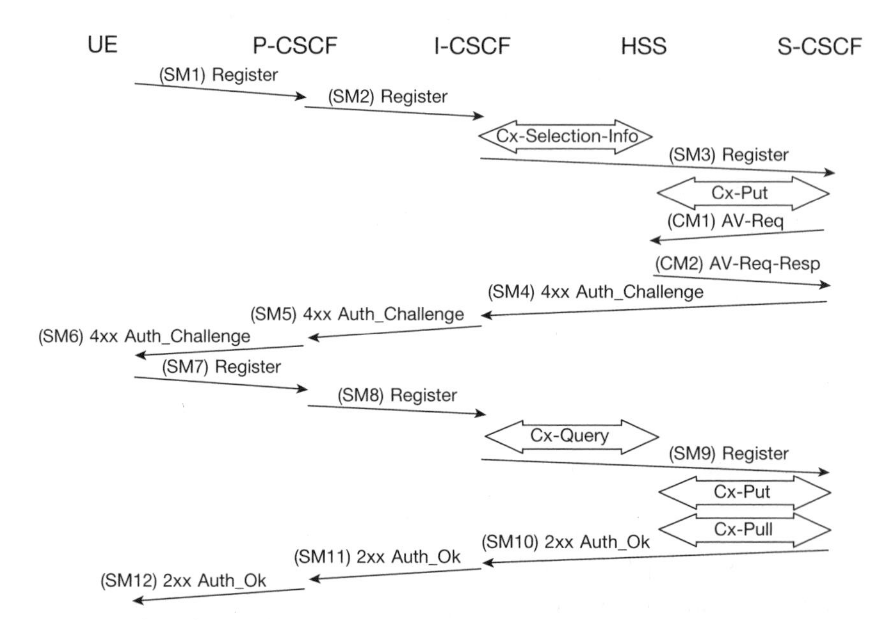

Figure 3.9 Registration and authentication in IMS

IMS = IP Multimedis CN Subsystem; IP = Internet Protocol; CN = Core Network; UE = User Equipment; CSCF = Call Session Control Function; HSS = Home Subscriber Server

security module on the user side (i.e., to reuse USIM [Universal Subscriber Identity Module]) as ISIM. In this case IMS-specific parameters have to be stored in the mobile terminal and there are also implications on the network side—most notably the HSS has to be able to utilize the same AuC that is used for the CS and PS domains (for this particular user).

At the same time as fetching the AVs, the HSS stores the address of the selected S-CSCF. Now, the S-CSCF picks up the first AV and sends three or four parameters (excluding XRES [expected response] and possibly CK [Cipher Key]) to P-CSCF via the I-CSCF. The P-CSCF extracts IK [Integrity Key] but forwards RAND (a random number) and AUTN (authentication token) to the UE. The SIP message used to carry all this information is *401 Unauthorized.* So, from a pure SIP perspective the first registration attempt has failed.

Nevertheless, the ISIM in the UE is now able to check the validity of AUTN, and (if the result of the check is positive) RES and IK are also computed. A parameter derived from RES is included in a new REGISTER request that is already integrity-protected by IK. Integrity protection is done by means of the IPsec ESP protocol. The IK is the most important part of the ESP SA in use.

The new REGISTER goes first to the P-CSCF and then is forwarded to the I-CSCF, which checks the validity of the S-CSCF address with the HSS. Note that the I-CSCF maintains no state for subscribers. The REGISTER is then forwarded to the S-CSCF, which then compares the parameter derived from RES with a respective parameter derived from XRES. If these two match, the *OK* message is sent all the way back to the UE.

The AKA procedure is now finished and the end result is as follows:

- the UE and P-CSCF share IPsec ESP SAs that can be used to protect all further communication between them;
- the S-CSCF and HSS have both changed the status of the subscriber from "unregistered" to "registered".

The registration and authentication procedure is depicted in Figure 3.8.

The S-CSCF always puts an AKA in place at the time of initial registration. For reregistrations, authentication may be skipped, depending on the choice of S-CSCF. It is also possible for the S-CSCF to force a reregistration of the UE at any time and, therefore, the S-CSCF may authenticate the UE whenever it wants.

All IMS security features will be discussed in more detail in subsequent sections of this chapter.

3.2.5 Use of HTTP Digest AKA

As already mentioned, SIP is an IETF protocol. However, 3GPP IMS needs its own extensions, which somehow have to be embedded in SIP. One such extension is the use of the 3GPP AKA as a mutual authentication and key agreement. A specific RFC 3310 [122] has been devoted to this issue, which builds on top of HTTP Digest [117].

HTTP Digest specifies headers that can be used for authentication of users. When using Digest it is assumed that the client and the server share a secret password. The idea is to verify on the server side that the client knows the password in such a way that the password itself is never sent over the network. Another scheme, called HTTP Basic [117], preceded HTTP Digest, according to which the password is sent over *as is*. Obviously, an attacker can eavesdrop the connection and pick up the password. This is why SIP no longer sanctions the use of HTTP Basic in its current specification [120].

The 3GPP AKA is not based on a password that is shared by the UE and P-CSCF; instead, the permanent secret is shared between the UE and the home network database server HSS.

In retrospect, there were at least three potential ways of getting an AKA introduced in SIP:

1. define new headers for the purpose of running 3GPP AKA in the HTTP framework (this could also have applied to SIP);

2. define a method in which an abstract authentication protocol like Extensible Authentication Protocol (EAP) [101] could be embedded in the HTTP framework (applicable to SIP) and, on the other hand, define a method of using the 3GPP AKA with this abstract protocol;

3. define a method in which the 3GPP AKA could be embedded in the HTTP Digest.

All these approaches have pros and cons. One important factor that had to be taken into account was time pressure: the freezing date of 3GPP release 5 was approaching fast. Eventually, the third alternative was chosen. The biggest problem with this approach has already been mentioned: the 3GPP AKA is not a password-based authentication. However, when Digest headers are defined in an appropriate manner, 3GPP AKA can be run with it. The main idea is to use RAND with AUTN as *nonce* and RES calculated from RAND as the password, which means that RES cannot be sent as such to the server (because in Digest the password is never sent as is). Instead, a one-way hash value, calculated based on RES, is sent over.

The headers defined in HTTP Digest [117] are called the *WWW-Authenticate* response header, the *Authorization* request header and the *Authentication-Info* header. The most important parameters (from the IMS point of view) in these headers are explained in the following:

- The *WWW-Authenticate* response header is included in the *401 Unauthorized* message that is sent as a response to an initial authorization request from

the user (REGISTER message in IMS). The header contains (a
things):

- ○ a *realm* directive that gives information about the host and the usei
authenticated;
- ○ a *nonce* directive that contains a base-64-coded data string (see RFC 204
[97], which is newly generated every time the header is created—it is used as a challenge to the user;
- ○ an *algorithm* directive that indicates how the user should compute the response to the challenge.

- The *Authorization* request header is then included in the new authorization request and contains, among other things:
  - ○ a *response* of 128 bits that proves the user knows the password;
  - ○ the *username* of the user in the specified realm;
  - ○ an *auth-param* directive that allows for future extensions.

RFC 3310 calls for UMTS AKA parameters to be mapped onto the HTTP Digest. AKA use is indicated by the *algorithm* directive, the *nonce* directive is populated by parameters RAND and AUTN, while RES is used as the password. In case of a synchronization error, the AKA provides a *resynchronization* procedure by means of the AUTS parameter (see Section 2.1.1.5), brought about in RFC 3310 by a novel *auts* directive.

In Figure 3.10 we show the information flow of a successful authentication.

If the authentication fails because of synchronization failure in the sequence numbers, the message flow is as described in Figure 3.11.

The HTTP Digest AKA does not specify how the IK and CK should be used; this is beyond the scope of RFC 3310. However, these keys are needed for protection of IMS first-hop signalling. As specified in [4] and [5], the S-CSCF appends the IK (and optionally the CK) to the *auth-param* directive in the *WWW-Authenticate* header of the *401 Unauthorized* message. The P-CSCF strips these parameters from the header and stores them for future use in first-hop protection.

3.2.5.1 Differences between the Digest AKA and a plain AKA

In the remainder of this section we explain how such a slight modification to the usage of a plain AKA imposed by the HTTP Digest AKA leaves no room for specific attacks (i.e., attacks that would be effective against [122] but not against the AKA itself).

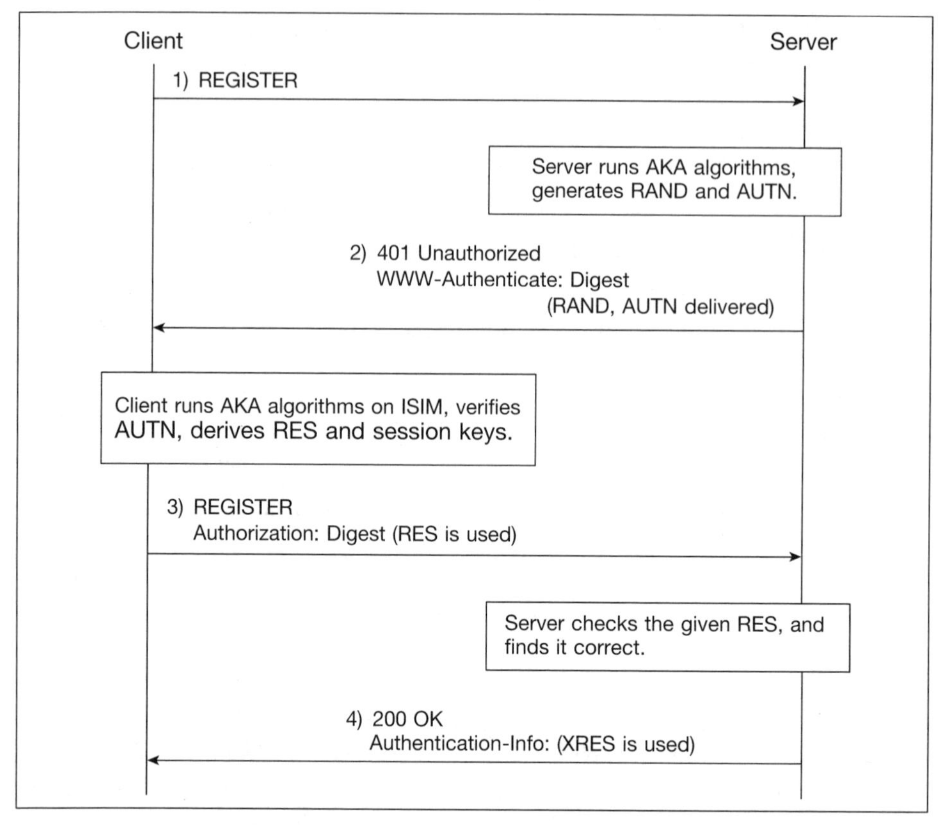

Figure 3.10 Message flow representing a successful authentication
AKA = Authentication and Key Agreement; RAND = random number; AUTN = authentication token; ISIM = IMS Subscriber Identity Module; IMS = IP Multimedia CN Subsystem; RES = user response; XRES = expected response

First, note that secret-key related data in the messages of [122] (i.e., MD5 [Message Digest #5] hash values) do not offer any clue to discovering any of the secret parameters in AKA. This is simply because in a plain 3GPP AKA the RES becomes public and any attacker can compute the same information (i.e., the MD5 values of messages in [122]) for himself.

Second, authentication of the network carried out by the UE demonstrates no difference between a plain AKA and the Digest AKA, since authentication can be done even before RES is created in the UE side. In other words, network authentication is based on verification of RAND and AUTN, which are sent in plaintext in both cases.

Third, when authentication of the user is carried out by the network, the basic question is: "for an attacker who does not know the master key K, is the probability of getting a response message accepted by the network greater in the Digest AKA case than in the plain AKA case?"

In both cases the attacker can try to guess the value of RES and prepare the

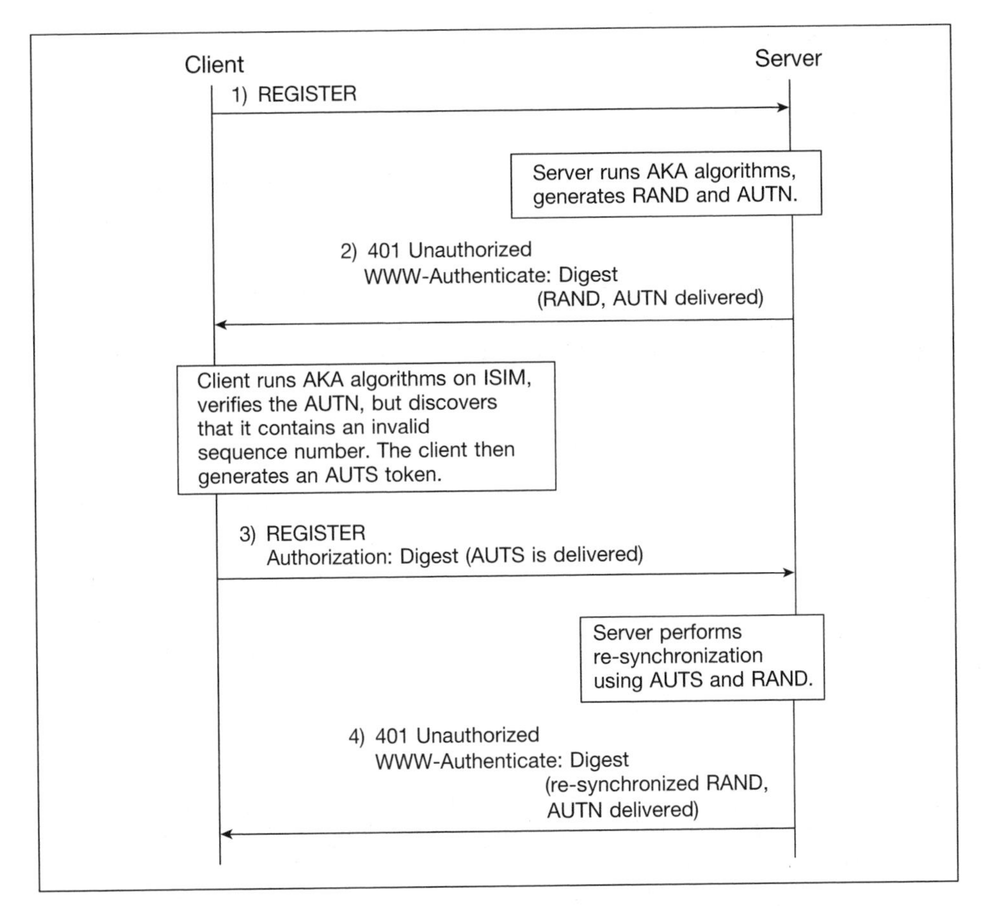

Figure 3.11 Message flow representing authentication synchronization failure
AKA = Authentication and Key Agreement; RAND = random number; AUTN = authentication token; ISIM = IMS Subscriber Identity Module; AUTS = synchronization parameter

response message accordingly. If the length of RES is n, then the probability of success is $1/2^n$.

The network server accepts authentication in the Digest AKA case only if the MD5 hash value in the response message is the same as the corresponding hash value calculated in the server using XRES as the password. In principle, it can happen that there are two XRES values that yield the same MD5 hash value. In this case, the attacker could get a response message accepted with a slightly better probability than guessing the RES directly. The problem, however, is that the attacker must first find out which MD5 value results from two different XRES values. In other words, the attacker has to find a specific type of MD5 *collision.*

The MD5 algorithm is not recommended for generic use as a one-way function, since there is a widely accepted suspicion that collisions could be found. However, it should be stressed that at the time of writing *no* actual MD5 collisions with *any* two inputs have yet appeared in the literature. Therefore, a major breakthrough in MD5 cryptanalysis has to happen before such an attack on the Digest AKA as described above is feasible.

Even in the (highly unlikely) event that a special type of MD5 collision can indeed be found, the probability of a successful attack is only doubled, and in any case it would be far easier to double the success probability of the attack by simply trying twice!

To improve the probability of a successful attack more dramatically, a joint MD5 collision of many input messages must be found. Even with a major breakthrough in MD5 cryptanalysis, this does not seem possible.

3.2.6 Security mode set-up

When a communication channel is protected by security mechanisms, the critical point is the very start of protection:

- What mechanisms are activated?
- At what point in time does protection start in each direction?
- What parameters (e.g., keys or SAs) are activated?

An obvious attack against any security mechanism is to try and prevent execution of the mechanism in the first place; even the strongest mechanisms are useless if you do not turn them on. This attack, and other similar attacks, can be carried out by a *man-in-the-middle*.

A new standards track specification was called for by RFC 3329 [123] and was created with the aim of securing negotiation between security mechanisms and SIP parameters. The need for such a specification arose from IMS security work, but the RFC is more widely applicable. Extensions to SIP are indicated by *option tags* included in certain header fields (e.g., in the *Require* header). RFC 3329 defines a new option tag called *sec-agree*.

The basic idea is, first, to exchange *security capability lists* between the client and the server in an unprotected manner and then check the validity of the mechanism choice later when protection is turned on. The message flow is depicted in Figure 3.12.

In the first step, the client sends the list of those security mechanisms it supports to the server. In the second step, the server sends its supported security mechanisms and parameters. In step 3, the client can discover which the highest preference mechanism the two parties have in common. Security execution starts with this mechanism. In step 4, the client returns the list of mechanisms and parameters that it received previously in step 2. In the final step, the server verifies that the

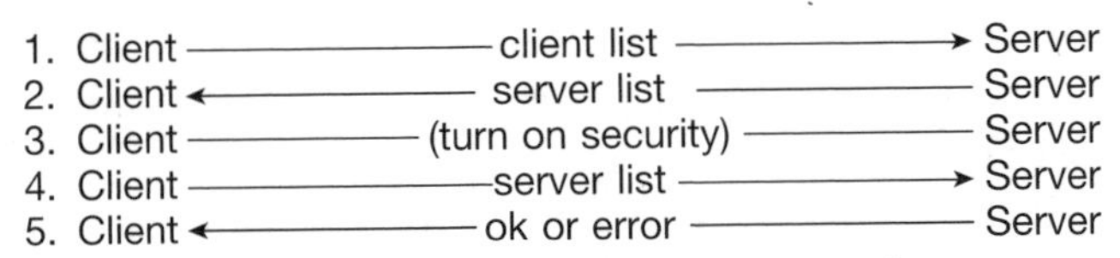

Figure 3.12 Security agreement message flow

list received from the client in step 4 is, indeed, identical to the list the server itself sent in step 2. Note that the mechanism allows the server to remain stateless. The client in the message flow above is the UAC, while the server could be the UAS, a proxy or a registrar.

Three new SIP headers were defined as a result of RFC 3329 [123]. These are called *Security-Client*, *Security-Server* and *Security-Verify*. In RFC 3329, there are four different security mechanisms that can be negotiated:

- *tls* for TLS [100];
- *digest* for HTTP Digest [117];
- *ipsec-ike* for IPsec used with IKE [103];
- *ipsec-man* for IPsec without IKE (but manually configured keys instead).

In addition to these four mechanisms, appendix A of RFC 3329 [123] defines a fifth:

- *ipsec-3gpp* for use of IPsec ESP to protect the IMS first hop.

A new response message *494 Security Agreement Required* is also required by RFC 3329 and is used to inform the client that the security agreement mechanism must be used.

3.2.7 Integrity protection with ESP

Mutual authentication is not sufficient to guarantee that, for example, all charging is made against the correct subscription, which is the reason that individual SIP signalling messages are also authenticated (note that they can be used to set up sessions).

Integrity protection (or message authentication) is done by the IPsec ESP protocol that was described in Section 3.1.2.3. Use of the IP layer protection mechanism implies some additional requirements. The identity against which messages are authenticated is the IP address in the case of ESP. On the other hand, charging related to IMS signalling is based on IMS identities, which are only visible at the SIP layer. Therefore, SIP layer identities, especially IMPI, have to be somehow tied to IP addresses. This problem is solved in IMS P-CSCF by checking the SIP layer, message by message, to see whether the IP address used for integrity protection is allowed for the IMPI in question. The link between the IP address and IMPI is originally created during the AKA (i.e., at the same time as the ESP SA is created).

It is unavoidable that some SIP messages have to be transmitted without integrity protection (e.g., the initial REGISTER or various error messages). In the IP layer the type of SIP message carried inside the IP packet cannot be seen which is why different *ports* indicating differentiation between unprotected and protected messages are used. Of course, it is essential to verify that no SIP messages that should be protected (according to the IMS security architecture) are actually sent through unprotected ports. To achieve this goal, the IPsec layer communicates information about the used IP address and port for each packet to the SIP application layer. The IPsec layer ensures the correct SA is used for each IP address and port, while the SIP layer ensures the IP address and port match the content of the SIP message (i.e., the SIP identities and the type of message).

The integrity protection algorithm (together with other parameters) is negotiated between the UE and P-CSCF as explained in the previous section. The AKA procedure results in the UE and P-CSCF sharing an IK, which is used to derive two pairs of unidirectional ESP SAs between the UE and P-CSCF. Two SAs are used for (both TCP and UDP (User Datagram Protocol)) traffic from the UE to the P-CSCF and the other two SAs are used in the opposite direction.

3.2.7.1 Security association parameters

The following parameters are included in the ESP SAs that are used for the IMS first hop:

- *an integrity protection algorithm*, which was negotiated between the UE and P-CSCF as a result of RFC 3329 (possible algorithms in Release 5 are HMAC-MD5-96 [105] and HMAC-SHA-1-96 [106]);
- an *SPI* (Security Parameter Index) is chosen by the UE for the SAs that are used in the P-CSCF to UE direction and by the P-CSCF for the SAs used in the opposite direction (SPIs are also negotiated by RFC 3329);
- an *SA duration* that always has the constant value $2^{32}-1$ seconds in the IP layer, but the lifetime of the SA is controlled by the SIP layer;
- a *mode* that is always transport mode;
- a *key length* of the integrity key IK_{ESP}—is 128 bits for HMAC-MD5-96 and 160 bits for HMAC-SHA-1-96.

The destination IP address and SPI together uniquely identify an SA in the IPsec layer both in the UE and in the P-CSCF. All SAs are stored in the SAD.

In addition to the SAD, an SPD is needed, which is used to decide whether protection is required for each outgoing or incoming packet. For the purposes of the SPD, SAs are bound to various *selectors*:

- source and destination IP addresses;
- transport protocols that may be used with the SA (in IMS, both UDP and TCP are allowed for each SA);
- source and destination ports.

Ports are used to differentiate protected messages from unprotected messages and to differentiate of new SAs from old ones in situations where SAs are renewed. In principle, the use of ports can allow cases where the same IP address is shared between several users and differentiation between them can be done with port numbers.

The P-CSCF picks two port numbers as the *protected* ports (port_pc and port_ps), which are communicated securely to the UE by RFC 3329 and are different from the standard SIP port number 5060. Only protected messages can be received by the protected ports at P-CSCF and it is the IPsec layer that takes care of this rule.

The UE on the other hand picks two local protected ports where only protected messages can be received (port_uc and port_us). At the UE, these port numbers are also used for sending protected messages. The numbers are communicated by RFC 3329. The protected client ports port_uc and port_pc are changed every time a new SA is created, but the protected server ports port_us and port_ps remain unchanged.

The use of port numbers for protected messages is summarized in Table 3.3. Unprotected messages may be sent and received on any port except the protected ports.

While the IPsec layer checks that all messages received by protected ports are indeed protected, it cannot guarantee that some messages that should have been protected were instead sent to unprotected ports. This checking must be done in the SIP layer. The only unprotected messages that the P-CSCF can receive are

Table 3.3 Use of protected ports for protected messages with UDP and TCP

UE port ↔ P-CSCF port	SIP requests (UDP and TCP) SIP responses (UDP)	SIP responses (TCP)
Uplink	port_uc → port_ps	port_us → port_pc
Downlink	port_us ← port_pc	port_uc ← port_ps

UDP = User Datagram Protocol; TCP = Transmission Control Protocol; UE = User Equipment; P-CSCF = Proxy Call Session Control Function; SIP = Session Initiation Protocol

REGISTER requests, while the UE can only receive responses to these unprotected REGISTER requests and other similar error messages without protection.

3.2.7.2 Management of SAs at the SIP layer

SAs are essential for the IPsec. Still, the connection between used SAs and the SIP identities can only be done at the SIP layer. At the P-CSCF, the SIP layer maintains a database where each SA is identified by the UE's IP address and the protected port number of the UE and the P-CSCF. Moreover, the corresponding SIP identities IMPI and IMPU(s) are listed for each SA. The lifetime of the SA is also recorded in the database.

For each incoming message at the P-CSCF, the SIP layer (after checking that protection was used in case the message is not a REGISTER request) checks that the SA used by IPsec matches the SIP identities inside the message.

At the UE, the database is simpler: for each SA, the corresponding protected port numbers are stored together with the lifetime and every incoming message is still checked against the database.

When the UE starts a reregistration, the request may be protected by an existing SA. In this case it is possible that no AKA protocol is done (i.e., the UE is not explicitly challenged). In case the AKA protocol has been executed, two new pairs of SAs are also created as a consequence. For some time during the registration, both the old SAs and the new SAs need to be stored. Indeed, the UE deletes the old *outbound* SAs after receiving the last (authentication successful) message of the registration procedure protected by a new *inbound* SA. The UE has to maintain the old *inbound* SAs until a further SIP message is received from the P-CSCF protected by a new *inbound* SA. This is due to the fact that the P-CSCF does not know for sure that the last message has been received by the UE until a further message protected by a new *outbound* SA is received from the UE. In addition to these rules, all SAs are naturally deleted whenever they expire.

At the P-CSCF, an additional complication is caused by the fact that the UE may start a new reregistration process while the previous reregistration process is still ongoing. This may happen, for instance, if the final authentication successful message never reaches the UE. In this case, the P-CSCF can have *six* SA pairs simultaneously (for the same UE). However, this exceptional case lasts for a very limited time before at least two of the six SA pairs are deleted.

3.2.8 Error case handling

There are several ways in which security mechanisms can fail, but there are no uniform ways of handling these error cases. In the following we list the most typical error situations.

3.2.8.1 Integrity check failure

As a general security principle, messages with incorrect integrity check values should be discarded without any further notifications. If the other party is a genuine user who just happens to use a wrong (perhaps old) key, then it will start from square one when it becomes apparent that there is something wrong with the integrity mechanism.

3.2.8.2 Authentication failures

There are four different causes of authentication failure:

- If user authentication fails (i.e., RES does not coincide with XRES), then the IK used to protect message SM7 would normally be incorrect as well and, therefore, integrity check failure would occur even before user authentication failure is noticed at the S-CSCF. If, however, the integrity check of SM7 passes, then the registration of the IMPI is cancelled from the HSS and the authentication failure response is sent back to the UE.
- If network authentication fails because the MAC in AUTN is not correct, then this state of affairs is indicated to the network in SM7. Again, in this case the IMPI is deregistered from the HSS.
- If network authentication fails because the SQN carried by the AUTN is not acceptable to the UE, then a resynchronization procedure follows. In SM7, the synchronization failure is indicated to the network and AUTS is included. The S-CSCF fetches fresh AVs from the HSS and the procedure essentially continues from the message SM4 again.
- If authentication fails because the S-CSCF does not receive any response from the UE within an acceptable time window, then the registration status of the IMPU in question is not changed from the state that existed before the authentication. Note that if the IMPU was already registered, the registration is not cancelled to avoid denial-of-service attacks against the users.

3.2.8.3 Errors in the security set-up

There are three different causes for security set-up failure:

- if the proposal of the UE in the message SM1 cannot be accepted by the P-CSCF, then an error response is sent back to the UE;

- if the proposal of the P-CSCF in the message SM6 cannot be accepted by the UE, then the UE simply terminates the registration procedure;

- if the *server list* returned by the UE in the protected message SM7 is not identical to the list sent by the P-CSCF in the message SM6, then the P-CSCF terminates the registration procedure.

3.3 Other Security Systems

As pointed out in the first chapter of this book, it is often the case that the same messages are protected by different security mechanisms in different layers. A typical example is where all PS domain data are encrypted with one key (CK) between the UE and RNC, but for banking applications there is another encryption algorithm in use between the UE and the server in the bank. It is usually not a good idea to try to reduce the number of layers to which security is applied because this kind of optimization would increase the risk that security is accidentally not applied in any layer. Bearing this in mind, we devote this last section of the present chapter to security mechanisms that are complementary or supplementary to the 3GPP security mechanisms presented so far.

3.3.1 Higher layer security systems

Although application layer security is not one of the key issues of this book, we take a brief look at the most common security mechanisms at higher layers.

3.3.1.1 Application layer security

At the application layer, popular security mechanisms include S-MIME (Secured Multipurpose Internet Mail Extension) and PGP (Pretty Good Privacy). The former is a security protocol that adds digital signatures and encryption to Internet MIME messages. It was originally developed by RSA Data Security Inc., who based it on triple-DES encryption and X.509 digital certificates. S-MIME uses the RSA public-key encryption method and the Diffie–Hellman system for key management. SHA-1 (Secure Hash Algorithm #1) is adopted for data integrity protection purposes.

PGP is freeware for email security originally designed by Philip Zimmermann. It brings privacy and authentication to email by using encryption and digital signatures. PGP uses IDEA (International Data Encryption Algorithm) for encryption and RSA for key management and digital signatures. Data integrity is protected by the MD5 algorithm.

One of the most interesting aspects of PGP is its distributed approach to key management. There is no Certification Authority (CA); instead, every user generates and distributes her own public key. Users sign each other's public keys, creating an interconnected community of PGP users. The benefit of this mechanism is that there is no CA that everyone has to trust. Each user keeps a collection of signed public keys in a file called a public-key ring. Each key in the ring has a key legitimacy field that indicates the degree to which the particular user trusts the validity of the key. The user sets this field manually. The weakest link of the PGP is key revocation. If someone's private key is compromised (e.g., stolen), a key revocation certificate has to be sent out. Unfortunately, there is no guarantee that everyone who uses the public key that corresponds to the compromised private key receives the key revocation in time.

3.3.1.2 Security for the session layer

At the session layer of the OSI (Open Systems Interconnection) stack both SSL (Secure Socket Layer) or TLS (Transport Layer Security) can be used. The SSL was originally developed by Netscape Communications Corporation to provide privacy and reliability between two communicating applications at the Internet session layer. SSL uses public-key encryption to exchange a session key between the client and the server. This session key is used to encrypt the HTTP transaction. Each transaction uses a different session key. Even if someone manages to decrypt a transaction the session itself is still secure (just the one transaction is violated). In the past encryption made use of a 40-bit (rest of the world) or 128-bit (USA) secret key, but the situation changes as export restrictions are relaxed.

3.3.1.3 AAA mechanisms

RADIUS (Remote Authentication Dial-In User Service) is described in RFC 2138 [99] as a protocol for carrying authentication, authorization and configuration information between NASs (Network Access Servers) and a shared authentication server. It provides a method that allows multiple dial-in NAS devices to share a common authentication database. It was originally developed by Livingston Incorporated for their Portmaster line of NAS products and has been widely deployed in many vendors' products over recent years.

The IETF AAA (Authorization, Authentication and Accounting) Working Group has developed a new protocol called Diameter. Diameter is used in the 3GPP Release 5 specification set for the Cx-interface between S-CSCF and HSS, and should gradually replace RADIUS as the dominant AAA mechanism.

3.3.2 Link layer security systems

Two major, wireless, link layer technologies based on radio frequency are Bluetooth [42] and IEEE 802.11 [58]. Dedicated security mechanisms for providing point-to-point security between the end points of the wireless link have been developed and specified for Bluetooth wireless communications and IEEE 802.11 wireless LANs. The mechanisms are end point authentication and link encryption. In addition, the Bluetooth system provides a secure pairing mechanism for establishing an initial link layer SA. IEEE 802.11 does not have such a system, but SA establishment depends on a higher layer authentication and key exchange mechanism, which most typically is provided by some EAP method [101]. In Bluetooth the link key can again be established at a higher communication layer and then transported to the Bluetooth module.

3.3.2.1 Wireless LAN and the EAP

The original IEEE 802.11 WLAN [58] security mechanism is known as Wired Equivalent Privacy (WEP). To access a WEP-protected network, a user must know the identity of the serving network, SSID (Service Set Identity), and the shared WEP key. Network identity is usually broadcast at the access point, so that the user can select it from a list. In many implementations the shared WEP key is the same key for all stations. WEP makes use of the RC4 stream cipher to provide authentication and encryption. WEP security has many serious problems (e.g., integrity and replay protection is missing and it is possible to use the authentication procedure to break the decryption). Moreover, the RC4 stream cipher is used with a short initialization value, which would require frequent rekeying. It is for this reason that VPN solutions are recommended for use in WEP-protected WLANs.

One of the failings of the WEP link layer security architecture is the absence of initial user authentication and link key establishment. The practical solution was to use the same key for all users. To improve the situation, the IEEE 802.1X authentication and key management framework [59] has been developed to enable authentication of individual devices and to distribute WEP keys. This framework makes use of the EAP [101], which is a standard interface for any authentication protocol. Using EAP, devices can be authenticated by means of passwords or public keys.

In 2002, project 802.11i began development of new, adequate security mechanisms for WLANs [60]. While still under development, a part of the future 802.11i was published in late 2002 in the specification of Wi-Fi Protected Access (WPA) endorsed by the Wi-Fi Alliance, which it should be noted is not an IEEE standard. Some vendors have developed similar but incompatible, proprietary products. WPA uses the 802.1X authentication and key management functionality, but replaces the

WEP encryption with TKIP (Temporal Key Integrity Protocol), which changes the way RC4 keys are used and adds message integrity and replay protection. In addition to 802.1X, WPA supports a Pre Shared Key (PSK) mode that allows manually entered keys or passwords. This mode is intended for use in standalone, personal area networks without access to an external key management facility.

The complete IEEE 802.11i, currently called the Robust Security Network (RSN), is still under development and not expected to be available until 2004. The main change is that the RC4-based TKIP will be replaced by a CBC-MAC (CCM) Protocol that uses the AES block cipher standard [90] for encryption and integrity protection of the link traffic. The standard may also include the Wireless Robust Authentication Protocol (WRAP), which is a similar protocol that also uses AES. RSN will also add several other enhancements. Previous WLAN security solutions have only supported Basic Service Set (BSS) mode (i.e., a network controlled by a single access point). RSN will also provide security for two other types of network configurations: Independent Basic Service Set (IBSS) and Extended Service Set (ESS). IBSS is a standalone network without a network access point (such as a personal area network) and ESS is formed by a set of multiple access points with a MAC layer handover and access point roaming functionality.

3.3.2.2 Bluetooth

The Bluetooth system has been developed by the Bluetooth Special Interest Group (Bluetooth SIG) as a cable replacement for short-range connectivity. The security mechanisms for the Bluetooth system are defined in the Bluetooth Baseband specification [42] and are based on strong cryptographic algorithms and well-founded security principles. The wireless link technology sets its limitations to the efficiency of the system. Bluetooth Baseband security is implemented at the Bluetooth module and is common to all Bluetooth units.

The Bluetooth link layer SA is most typically established using the Bluetooth pairing procedure, in which a 128-bit shared secret key, a *link key*, is created between two Bluetooth units. Two types of link keys are specified: *unit keys* and *combination keys*. The specification allows the link key to be established at a higher communication layer of the device and then imported to the Bluetooth unit using the Host Controller Interface (HCI).

The link key is used to authenticate Bluetooth units to other Bluetooth units. During the authentication process one unit, the verifier, sends a random value to the other unit, the claimant. The claimant has to process the random value together with the link key to obtain a correct response value. The response value is sent back to the verifier who compares the received value with an expected value precalculated by the verifier. The authentication only works one-way and if units want mutual authentication, two consecutive authentication processes must be performed. As a side result, the authentication process generates a bit string, the Authentication Ciphering Offset (ACO), which is used for ciphering key generation. The ciphering key is calculated as a cryptographic hash of the link key, a random value and the ACO. The cryptographic functions used for authentication and key derivation in Bluetooth are based on a block cipher algorithm (SAFER++), which was one of the AES round 1 candidates [88].

The Bluetooth encryption algorithm E_0 is a stream cipher (keystream generator) with a 128-bit key and a 128-bit internal state. E_0 was developed by the Bluetooth specification group and was published in 1999 as a part of the Bluetooth Baseband specification. Since then it has been analysed by a number of well-known cryptanalysts and their findings have been published. Such a stream cipher (i.e., with a 128-bit internal state) is likely to be broken with a workload of about 2^{64} steps, given a keystream length of 2^{64}. Since it is more difficult to collect data than to make offline computations, the target of cryptanalysis is to improve this universal attack by trading data for computational complexity. One of the latest papers [73] presents the best known practical attack so far: using a short 128-bit keystream and carrying out about 2^{73} computation steps it is possible to derive the internal state of the algorithm. This is comparable with the expected complexity of an attack which performs an exhaustive key search for a cipher with a 74-bit key. Under current knowledge, this means the effective key length (Section 4.2.1.3) of the Bluetooth stream cipher is 74 bits, thus setting a theoretical bound to the level of Bluetooth security.

Part II

Cryptographic Algorithms

4

Introduction to Cryptography

4.1 The Science of Cryptology

4.1.1 Cryptographic systems

Cryptology is the science of information security and privacy. Mathematical techniques are investigated and developed to provide authenticity, confidentiality, integrity and other security services to information that is communicated, stored or processed in an information system. The strength of cryptographic designs and protocols are evaluated from the point of view of mathematics, systems theory and complexity theory.

The design part of the science is called cryptography, while the security investigations and analysis are known as cryptanalysis. The naming convention reflects the two sides of the science of cryptology, a division that is also apparent in practical cryptographic development work, where best practice splits development resources into two teams: a team of cryptographers make proposals for cryptographic designs that a team of cryptanalysts try to break.

A cryptographic system in its basic form is often depicted as a communication system involving three entities. Two of these entities exchange messages over an insecure communication channel. It has become customary to call these entities Alice and Bob. The third entity has access to the communication channel. She is called Carol, as the third letter to the alphabet, or Eve, as the eavesdropper. But Eve is allowed to perform all kinds of malicious actions on the communicated messages, not just passive eavesdropping. All parties are also assumed to have certain computation resources. Different theoretical models vary a lot with respect to the amount of computation resources the entities have and what kind of tampering Eve carries out on the communication channel.

The goal of cryptography is to ensure that communication between Alice and Bob is secure over an insecure channel. A cryptographic system is typically given as a family of cryptographic functions, parameterized using a cryptographic value called the key. The functions may be invertible or non-invertible. Invertible functions are

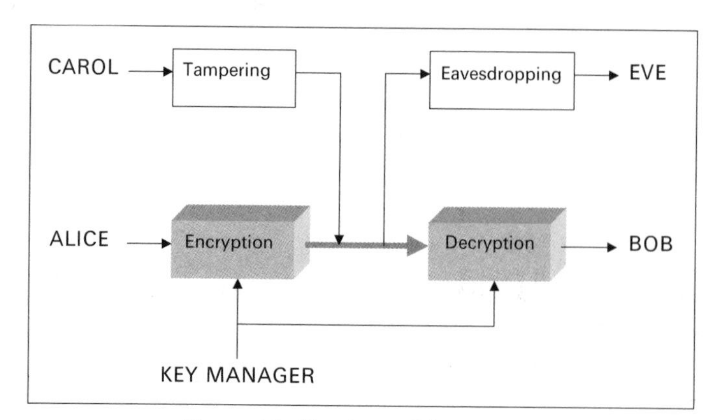

Figure 4.1 Cryptographic system

needed to protect the confidentiality of the messages. The message (plaintext) is encrypted by the sender entity (Alice) using the function. The encrypted message (ciphertext) is then sent over the channel to the receiving entity (Bob). Bob decrypts the ciphertext using the inverse function. Non-invertible functions are only computed in one direction and are useful in protecting the integrity of messages. Examples of cryptographic systems that use non-invertible functions are message authentication codes (see Section 4.2.3).

Description of the cryptographic system can be made public, and even known to Eve. The security of a cryptographic system does not depend on the secrecy of the system. Hence cryptographic algorithms can be published, distributed and sold as commercial products. The users of the cryptographic system, Alice and Bob, are only required to keep secret the knowledge of the actual function they are using. They indicate their selection to the system by giving the system the key, the value of the cryptographic parameter. To outsiders (Eve and Carol) the selection of the shared secret of Alice and Bob must be unpredictable to provide full uncertainty of the function Alice and Bob are using. Hence there is no secrecy without uncertainty. Uncertainty is created by randomness. Cryptology investigates how randomness can be efficiently used to protect information. The main challenge to the management of cryptographic keys is to provide unpredictable keys to users of the cryptosystem. The requirement of unpredictability has often been underestimated in practice, or has been traded off for other requirements. Cryptographic keys using *aides-memoires* (memorable poems or items of literature) have often turned out to be disastrous. A lively description, entitled *Between Silk and Cyanide*, of the various aspects involved in the generation, management and use of cryptographic keys is given by Leo Marks in [79].

The science of cryptanalysis has identified a number of ways in which Eve or Carol could attack the cryptosystem. Of course, the goals of the attacks can vary:

Eve may just want to eavesdrop, while Carol may want to forge messages and eventually create a "triangle" [44]. Eve may use ciphertext alone, while Carol may use a specific chosen ciphertext. The ultimate goal is to find the secret key, since it would mean a total breakdown of Alice and Bob's cryptographic system. However, the occasional compromise of one key used by Alice and Bob would not ruin the system: Alice and Bob would just need to change to a new key and take better care of it. A cryptosystem is considered to have been totally broken if there exists an efficient method by means of which the key can be systematically derived from practically-available information with a non-negligible probability of being found out.

4.1.2 Security and vulnerability

Before the revolution in information technology as a result of computers (e.g., during World War II), professional research and development of cryptography was a privileged activity of military and intelligence organizations. Primer-level textbooks were published explaining the basic principles of classical cryptosystems and their cryptanalysis. Such material is still often included as an introductory chapter in modern cryptographic textbooks [131]. Cryptology of a more serious nature was a carefully-protected proprietary knowledge. The first trustworthy and detailed accounts about the cryptanalysis of the German Enigma cipher machine were not published until the late 1970s. The extensive British cryptanalytic activity during World War II that took place in Bletchley Park, a small village between the university towns of Oxford and Cambridge, remained a well-kept secret for 30 years after this activity ceased at the end of the war.

Cryptographic technology has not lost its importance for national security, nor have intelligence organizations lost their interest in cryptologic research. Cryptographic methods and the devices carrying such methods are considered articles of warfare and, therefore, their use and export is controlled. The governments of 33 industrial countries participate in the Wassenaar Arrangement on export controls for conventional arms and dual-use of goods and technologies [134]. The purpose of this arrangement is to prevent proliferation of weapons of mass destruction technology to governments of less stable countries. The power of cryptographic systems is measured by their key sizes. There has been some more pressure to increase control of the use of cryptography in the aftermath of September 11th, where it has been claimed that the attackers used encrypted email for their communication. The *New York Times* interviewed inventors of modern cryptography, among others Stanford professor M. E. Hellman, asking such impossible questions as: what would have happened had they refrained from publishing their inventions? (most of which dated back to the 1970s) or did they know then to what extent their inventions would be misused? [72]. In this respect, cryptography shares the dilemma that plagues much of modern technology: how to exploit the best of it without creating new vulnerabilities.

4.1.3 Developing cryptology into a publicly available science

The first, modern, scientific treatment of cryptology was published in 1949. It was a comprehensive paper by Claude Shannon that presented the theoretical framework of what he called secrecy systems [130]. A paper on the mathematical theory of communication [129], published the year before, was a seminal work that laid the foundations of modern information theory in terms of bits and was the catalyst for the growth and successful research activity in this field soon after it was published, but did not result in any upsurge in *open* cryptologic research. Indeed, it took 27 years for anything significant to happen. James Massey considered the reasons for this in his very readable survey paper on the principles of contemporary cryptology [80]:

> First, the theory of theoretical security of secrecy systems that it provided was virtually complete in itself, and showed conclusively that theoretically-secure secrecy systems demand the secure transfer of far more secret key than is generally practicable. Moreover, the insights that Shannon provided into the practical security of secrecy systems tended to reinforce accepted cryptographic approaches rather than to suggest new ones. But Shannon's observation that "The problem of good cipher design is essentially one of finding difficult problems, subject to certain other conditions. ... We may construct our cipher in such a way that breaking it is equivalent to (or requires at some point in the process) the solution of some problems known to be laborious" took root in the fertile imaginations of Stanford cryptologic researchers, W. Diffie and M. E. Hellman. The fruit was their 1976 paper "New Directions in Cryptography" that stunned the cryptologic world with the startling news that *practically-secure secrecy systems can be built that require no secure transfer of any secret key whatsoever* [author's italics].

So, there was no public research activity for more than a quarter of a century, while within *closed* organizations the research was continued by experts, developing encryption machinery and analysing wiretapped, encrypted, communication traffic. In most countries the use of cryptography and cryptographic equipment was subject to license and limited to securing the internal communication of governments. During the Cold War, mathematicians developed encryption systems, but they were not made public. Results of mathematical research were withheld from publication if they were considered applicable to cryptography (e.g., O. Rothaus discovered mathematical objects he called "bent functions" in the 1960s, but the paper did not appear until 1976 [126]. In addition to encryption technology, bent functions have applications in spread spectrum technology, which was initially developed for military radio communication and later became the radio technology of UMTS (Universal Mobile Telecommunications System). The coding sequences that are used to effectively spread radio channels over the spectrum are based on bent functions and similar mathematical constructions. Even in the late 1970s researchers at the Massachusetts Institute of Technology (MIT) were forbidden to publish their results based on the export control of conventional arms [78].

In Shannon's secrecy systems, the enemy (i.e., the cryptanalyst) has access to the encrypted message and is also assumed to have detailed knowledge of the used cryptosystem that defines the family of cryptographic functions that constitute the cryptosystem. This principle is called "Kerchoff's principle" after a Dutch linguist A. Kerchoff. The secrecy of the system is based solely on the secrecy of the key.

Before security features can be implemented in large public systems, such as modern computer and communications networks, they must be thoroughly scrutinized using well-founded scientific and engineering principles.

Initial design efforts for a public cryptographic method were launched in 1973 in the USA when the National Bureau of Standards made an open call for an encryption algorithm suitable for data protection in commercial and banking communication networks and databases. It took four years before the Data Encryption Standard (DES) was published in January 1977. The DES is a conventional algorithm based on Shannon's principles and the cryptographic experience of IBM and NSA experts. Publication of the DES algorithm took place within a year of the publication of the revolutionary paper of Diffie and Hellman. These two events constituted the starting points of modern cryptologic research. Ever since, the DES algorithm has been an inexhaustible source of cryptologic research material in the field of symmetric (or conventional) cryptography, while the work of Diffie and Hellman opened up new and unconventional directions for public key cryptography.

While the scope of cryptologic research became wider, the range in the various security services provided by cryptographic applications started to grow rapidly from traditional protection-of-message confidentiality to authentication-of-communication entities as well as protection of data integrity. The formal concept of a cryptographic one-way function was created. The first-known examples of practical systems using one-way functions for authentication purposes date back to the 1950s. These Identification Friend or Foe (IFF) methods were used for authentication between military aircrafts [49]. In the early 1970s the first applications of one-way functions, although not called by that name at the time, were made to protect password tables in computer servers [135]. Essentially the same paradigm is used by the GSM (Global System for Mobile) and UMTS specifications for subscriber authentication in modern mobile communication systems.

Scientific work in cryptologic research is active and successful. The International Association of Cryptologic Research (IACR), founded in 1982, organizes three major conferences each year in the USA, Europe and Asia or Australia. In addition it helps to organize smaller, specialized workshops, such as the annual workshop on fast software encryption, and publishes a scientific journal, the *Journal of Cryptology*. The conference and workshop proceedings and the journal published by the IACR constitute the main body of scientific cryptologic literature.

4.1.4 Public cryptographic development efforts

In addition to scientific research, the public international development and research efforts contribute significantly to the general knowledge and understanding of the security and performance requirements of modern cryptographic systems. The development and analysis of cryptographic algorithms for the Third Generation Partnership Project (3GPP) UMTS benefited significantly from two such projects: the Advanced Encryption Standard (AES) programme by the National Institute of Standards (NIST) and the New European Schemes for Signatures, Integrity and Encryption (NESSIE) project.

The overall goal of the AES programme was to develop a Federal Information Processing Standard (FIPS) that specifies an encryption algorithm capable of protecting sensitive government information well into the 21st century. The algorithm was expected to be used by the US Government and, on a voluntary basis, by the private sector. The initial announcement of an open AES competition was published on January 2nd, 1997 and the AES development and evaluation process took four years (documentation is available at the AES home page [88]). After the first round, five candidates were selected for the second round, from which the Rijndael cipher of Belgian origin with its 128-bit block and three different key sizes was selected as the winner of the competition in September 2000. The 3GPP MILENAGE (see p. 215) algorithm makes use of the AES algorithm as its "cryptographic engine", as described in Chapter 8.

The NESSIE project is a three-year (2000–2003) project within the fifth framework of the European IST (Information Society Technologies) programme [86]:

> The main objective of the project is to put forward a portfolio of strong cryptographic primitives that has been obtained after an open call and been evaluated using a transparent and open process. The project intends to contribute to the final phase of the AES block cipher standardisation process (organised by NIST, US), but will also launch an independent open call for a broad set of primitives providing confidentiality, data integrity, and authentication. These primitives include block ciphers, stream ciphers, hash functions, MAC [Message Authentication Code] algorithms, digital signature schemes, and public-key encryption schemes. The project will develop an evaluation methodology (both for security and performance

evaluation) and a software toolbox to support the evaluation. The project goal is to widely disseminate the project results and to build consensus based on these results by using the appropriate fora (a project industry board, 5th Framework programme, and various standardisation bodies). A final objective is to maintain the strong position of European research while strengthening the position of European industry in cryptography.

The block cipher algorithm MISTY1 is one of the NESSIE candidates and has been extensively evaluated within the NESSIE project. Since the many attacks attempted on MISTY1 may also be relevant to 3GPP KASUMI and vice versa, the extensive analysis performed by the NESSIE project on MISTY1 has also consolidated the position of KASUMI as a secure cryptographic primitive.

4.2 Requirements and Analysis of Cryptographic Algorithms

The goal of this section is to give a brief introduction to the concepts of the most important types of cryptographic algorithms that make use of a secret key as the main security parameter. The basic structural, security and implementation features will be covered. A more complete presentation of the fundamentals of analysis and design of cryptographical algorithms can be found in the NESSIE report [87], where also several examples of typical, cryptanalytical attacks are presented (a comprehensive handbook of modern cryptographical concepts is [84]).

The most important cryptographic primitives from the point of view of UMTS security are the block and stream ciphers, used for confidentiality protection, and the message authentication codes, used for proving message integrity and often as a cryptographic primitive in entity authentication protocols. In the 3GPP UMTS security system, voice and data are encrypted using a stream cipher algorithm and integrity of the data is provided using a message authentication code. The protocol for mutual authentication of the subscriber and the network makes use of different types of message authentication codes, while key generation is a type of pseudorandom bit generator.

Cryptanalysis forms an essential part of the design process of a cryptographic algorithm. Its goal is to identify as many attacks as possible that can be launched against the algorithm. The strength of the algorithm is evaluated in terms of the complexity of the attacks. The design structure and the parameters of the algorithm are adjusted if necessary to make it infeasible in practice to launch any of the identified attacks. Hence, although it may sound paradoxical, the more attacks that are known and analysed for a cryptographic algorithm the more evidence we get of its security. This principle is known as "practical security" and will be discussed together with other important security notions in Section 4.2.1.2.

4.2.1 Block ciphers

4.2.1.1 Security requirements

A block cipher is an encryption algorithm defined as a family of invertible functions parameterized using the secret key. The block cipher partitions a plaintext message into a sequence of blocks of equal length, which are then taken as input for block cipher transformation, encryption or decryption, one block at a time. Many ideas introduced by Shannon [130] in 1949 are still valid and used in the design of modern block ciphers (DES and Lucifer are the first published examples). Shannon introduced the criteria of confusion and diffusion, which are still valid. New design criteria are being discovered, often as a response to new attacks on block ciphers, and when a state-of-the-art block cipher is constructed, all known attacks and developed design principles are taken into account. But no such block cipher can become absolutely secure but may remain open to some new, unforeseen attacks.

The block cipher transforms a plaintext block of length n to a ciphertext block of equal length n under the control of a secret key K. For each key, the transformation is invertible, so that ciphertext blocks can be decrypted using the decryption transformation. A block cipher with block length n and key size k bits constitutes 2^k invertible transformations (also called permutations), which is an infinitesimally small fraction of all $(2^n)!$ invertible transformations of blocks of n bits. The basic threat against confidentiality is that it is possible to retrieve the plaintext from the ciphertext without knowledge of the key. The basic security goal is to protect against this threat. In the theory of block ciphers it has become customary to define the security goals in terms of resistance against security threats or "breaks". Therefore it is important to identify the relevant threats posed to block ciphers (i.e., what it means to break a block cipher). The following list of breaks was presented by Lars Knudsen in [69]:

- *Total break.* An attacker finds the secret key K.
- *Global deduction.* An attacker finds an algorithm A, functionally equivalent to E_k, without knowing the key.
- *Instance deduction.* Given an intercepted ciphertext, an attacker finds the plaintext, which she did not obtain from the legitimate sender.
- *Information deduction.* An attacker gains some (Shannon) information about the secret key or the plaintext, which she did not get directly from the sender and which she did not have before the attack.
- *Distinguishing algorithm.* An attacker is able to tell whether the attacked cipher is a randomly chosen permutation or one of the 2^k permutations specified by the block cipher algorithm.

The attacks are listed in hierarchical order: from the most specific to the most generic. The more general the attacks it can withstand the stronger the block cipher. Hence if a block cipher could be proved to resist all attempts that try to distinguish its encryption or decryption transformation from a random permutation, then the block cipher would also be proven secure against all the other attacks listed above.

In his seminal paper [130] Shannon considered how much information a given ciphertext yields about the plaintext and the key. According to his definition a secrecy system provides *perfect secrecy* if the probability distribution of the plaintext is not changed, given the knowledge of the corresponding ciphertext, whatever the a priori plaintext probability distribution might be. In today's terminology it is then said that the cipher is "perfect". Shannon also proved a result that was called Shannon's Pessimistic Inequality, which states that, in order to achieve perfect secrecy, the key of the cipher must be at least as long as the plaintext that is encrypted (e.g., a block cipher with a 128-bit key could be used to encrypt at most 128 bits of plaintext, after which a new fresh key must be generated). The Vernam Cipher, also called a one-time pad, is a perfect cipher (it simply adds the key to the plaintext bit by bit). However, ciphers are usually far from being perfect in Shannon's sense, because in order to be practical a cipher uses a short string of key to provide confidentiality for a much longer string of plaintext. Therefore, other more practical security notions have been developed (e.g., the goal of the computational security model is that the length of the plaintext data to be encrypted using a single key is exponential to the length of the key).

4.2.1.2 Security model

The science of cryptology has developed different definitions of security for ciphers. Since practical ciphers do not fulfil Shannon's definition of perfect secrecy, other more practically-oriented definitions have been developed. Shannon's perfect secrecy means *unconditional security* in the sense that the attacker is allowed unlimited computational resources. The following definitions of security are most commonly used in practice:

- *Computational security*. This measure concerns the computational effort required to break a cipher. In the finite model of computational security, a cipher is said to be computationally secure if the best algorithm for breaking it requires at least N operations, where N is a sufficiently large number. In the asymptotic model, a cipher is said to be computationally secure, if the best algorithm for breaking it requires an amount of computational resources that increases faster than any polynomial function of the input size as the input size increases. The finite model is used to consider the security of a cipher of fixed size, while the asymptotic computational model is suitable for variable-sized constructions.

- *Practical security*. This measure also concerns the computational effort required to break a cipher, but now only with respect to previously known attacks. A cipher is said to be practically secure if the best known algorithm for breaking it requires at least N operations, where N is some fixed number, or requires an amount of computational resources that increases faster than *any* polynomial function of the input size as the input size increases. A closely related notion is that of *historical security*. The number of attacks against which the security of a cipher has been analysed does not decrease and hopefully increases over time. In this sense a cipher achieves historical security the more time it has been under scrutiny and the more effort that has been spent cryptanalysing it.

- *Provable security*. This means that the cipher has a proof of security under some assumptions. The proof is more valuable the more relevant the assumptions. Typical examples include the following:

  - Let us assume that a given computational problem is difficult and that complexity theory has identified several problems conjectured to be difficult. A cipher can be proven secure by showing that breaking it requires at least as many computational resources as solving the difficult computational problem.

  - The security of the cipher is considered with respect to a certain subclass of attacks, against which the cipher is proven to be secure.

  - The cipher is proven secure assuming that some functional part of it satisfies certain requirements.

These security notions can also be adapted to other types of cryptographic primitives, such as stream ciphers and MAC (Message Authentication Code) functions.

4.2.1.3 Classification of attacks

To measure the security of a computational model requires an estimation of the resources, time, memory and data the considered attack methods take to achieve the break.

One design goal of a block cipher is to use the secret key as efficiently as possible. The success of this goal is measured by comparing the time complexity of an attack with the time complexity of an exhaustive key search, which is equal to 2^k for a cipher with key length k bits. If an attack method is found, which leads to a break (in some relevant sense, see Knudsen's list in Section 4.2.1.1) in time 2^t, where $t < k$, then the cipher is said to be *theoretically broken* and the *effective key length* of the cipher is reduced to t. This does not necessarily mean that the cipher is broken in practice, unless t is small enough.

Such an exhaustive key search requires a very small amount of data and uses very little memory. However, the most-practical attacks require a significant amount of memory, which is usually considered more expensive than time. Data are also considered expensive and should be minimized. Various trade-offs are possible, but the complexity of the attack is usually taken as the maximum of the amounts of time and data.

Attack methods are traditionally classified according to the type of data available to the adversary as follows:

- *Ciphertext-only attack*. The attacker has access to an amount of ciphertext and knows something about the nature of the plaintext that is not perfectly random.
- *Known plaintext attack*. The attacker has access to an amount of plaintext and the corresponding ciphertext.
- *Chosen plaintext attack*. The attacker is able to choose an amount of plaintext and obtains the corresponding ciphertext.
- *Adaptively chosen plaintext attack*. The attacker is able to choose parts of an amount of plaintext and obtain the corresponding ciphertext, where the choice of each new part of plaintext is influenced by all previously obtained ciphertext.
- *Chosen ciphertext and adaptively chosen ciphertext attacks*. Similar to the chosen plaintext attacks, but now with the roles of plaintext and ciphertext reversed.

Modern cryptanalysis has developed a number of different attack methods. The most important attacks on block ciphers are made by differential and linear cryptanalysis methods, since they together with their variations apply to a large variety of different types of iterated block ciphers. For an overview of the most commonly used cryptanalysis methods of block ciphers, the reader is referred to [87], where many examples are presented. Some of the attacks used to analyse the MISTY1 and KASUMI block ciphers will be explained in Sections 7.2, 7.5 and 7.6.

4.2.1.4 Common designs and tools

All practical designs of block ciphers involve a small number of simple transformations that constitute one round of a block cipher. The most commonly-used simple transformations are bit permutations and small substitution transformations, the so-called S-boxes. Bit permutations are used to diffuse as efficiently as possible the influence that each input data bit and each key bit has over the whole block, while the main purpose of S-boxes is to hide as efficiently as possible all interrelationships between different bits. An iterated block cipher is constructed by repeating the round transformation a suitable number of times. Most block ciphers are either Feistel Ciphers or Substitution–Permutation Networks (SPNs). In a round function of the Feistel Cipher a substitution transformation is only applied to half the input data, therefore making the construction invertible, independently of whether the substitution transformation is invertible or not. In SPN, substitution transformation operates on all the input data at once, which means that it must be invertible.

Research into block ciphers has produced a number of useful building blocks and components for use in block cipher engineering. An insightful view to the design and analysis of contemporary block ciphers can be found in [69]. In Chapter 7 a detailed account of the design of the KASUMI block cipher is presented. The interested reader is encouraged to consult [69] and [87] for useful background information.

4.2.1.5 Modes of operation

In practice, block ciphers are very rarely used in basic memoryless mode, where encryption of each block is independent of the other data blocks. This mode, also known as Electronic Code Book (ECB) mode, has certain vulnerabilities. Because of this, it is common to embed the block cipher in a framework with memory. Different modes of operation have been developed in which block ciphers are used to provide different types of cryptographic algorithms such as block ciphers, stream ciphers, MAC algorithms and pseudorandom bit generators. The four most widely known modes of operation were originally standardized by NIST for use with the DES algorithm [89]. Probably the most commonly used mode for block encryption is Cipher Block Chaining (CBC) mode, where encryption of a plaintext block depends on the encryptions of previous blocks. The other two modes, the Cipher Feedback (CFB) and the Output Feedback (OFB), make block ciphers operate as stream ciphers. Another commonly used mode of operation is the counter mode, which is similar to OFB with the only difference that it involves updating the input block as a counter [84]. Counter mode is used to construct stream ciphers and pseudorandom bit generators. Block ciphers can also be used to construct MAC functions, such as CBC-MAC [61].

These modes of operation are widely used and their security has been closely analysed during "mock" cryptanalytic attacks. For instance, there is a straightforward ciphertext-only attack that allows the attacker to retrieve some information about the plaintext. The attack makes use of the fact that in a relatively small set of ciphertext blocks the attacker is likely to find two equal ciphertext blocks, often called a "collision". According to the Birthday Paradox, if the block size is n bits then the probability of finding a collision in a set of just over $2^{n/2}$ randomly chosen

n-bit blocks is about one-half (e.g., if such a collision of ciphertext blocks is found during CBC mode, the attacker gets to know the XOR (exclusive OR) of the corresponding plaintext blocks). In ECB mode, the distribution of ciphertext blocks is the same as the distribution of plaintext blocks, and hence a collision is likely to occur much faster.

The attack described above is a typical example of a theoretical attack that sets limits to the security of the system. Because the possibility of the above attack, it is not advisable to use the same key to encrypt more than about $2^{n/2}$ plaintext blocks. For practical systems, the block length n is defined to be large enough to ensure that this security limit is never met.

4.2.2 Stream ciphers

4.2.2.1 Introduction

A stream cipher is a cryptographic algorithm for encrypting plaintext. Similarly to block ciphers the plaintext is partitioned to a sequence of blocks. The main difference between stream ciphers and block ciphers is that in the former the current plaintext block is not taken as data input for cryptographic transformation. Instead, a string of bits, often called a "keystream" block, is generated independently of the current plaintext block and then combined with the plaintext using a simple operation, most commonly the XOR operation. Due to this functional difference, stream ciphers typically operate on plaintext blocks of shorter length, which can be just 1 bit or 1 byte. The collection of all possible plaintext blocks is often called the "plaintext alphabet".

At the decrypting end, the same keystream block is generated and XORed to the ciphertext block. In this manner the effect of the keystream is cancelled and the plaintext block is recovered. It follows that the algorithms for encryption and decryption are identical. The internal state is initialized using the secret key K and some other, possibly public, initial value. Then, for correct operation the internal states of the two keystream generators must be synchronized so that exactly the same keystream block is generated for the same data block. A stream cipher is called self-synchronizing if ciphertext is used as input to the state; otherwise, a stream cipher is called synchronous. Stream ciphers offer some advantages in specific applications [87]:

- stream ciphers are generally faster than block ciphers, especially when implemented in hardware;
- stream ciphers have less hardware complexity;

- stream ciphers can be adapted to process the plaintext bit by bit, or word by word, while block ciphers require buffering to accumulate the full plaintext block;
- synchronous stream ciphers have no error propagation.

For more information on the theory and practice of stream ciphers, see [84] and [127].

4.2.2.2 Classification of attacks

Stream ciphers do not use iterated functions in the same way as block ciphers. The state update function is iterated, but a small amount of output is produced at each iteration round, while block ciphers do not produce output between the rounds.

Considering the resources available to a cryptanalyst, it is most customary to assume a certain amount of known plaintext, which means, since the corresponding ciphertext is available, that a certain amount of the keystream is known to the cryptanalyst. Clearly, since plaintext is not taken as input to the cryptographic function, a chosen plaintext does not add anything to known plaintext, while a chosen ciphertext attack may be a useful method for analysing a self-synchronizing stream cipher.

The attacker of a stream cipher may try to break it in different ways: it is a straightforward procedure to adapt the breaks given for block ciphers in Section 4.2.1.3 to those for stream ciphers. In stream ciphers the "global deduction" attack and the "distinguishing" attack are of particular importance. In the former, the attacker only needs to find one internal state of the stream cipher to obtain a functionally equivalent algorithm without knowing the key. Distinguishing a keystream sequence from a truly random sequence also allows the keystream to be predicted with some accuracy (such an attack is also called a "prediction" attack).

Similarly to block ciphers, the security of all practical stream ciphers is computational. The only unconditionally secure stream cipher is the one-time pad.

4.2.2.3 Common designs and tools

One of the goals of the design is to ensure that, given a fixed initialization value, the stream cipher generates a different keystream for each secret key. Moreover, it is desired that the period of each generated keystream is maximal.

This requirement is satisfied by a simple device called a Linear Feedback Shift Register (LFSR). LFSRs are often used as the running engines for a stream cipher. An LFSR is defined by a simple linear transform which, if given by a primitive polynomial, provides a maximum possible cycle length. If the length of the LFSR state is k, then the maximum length of the cycle is $2^k - 1$. An LFSR has perfect statistical properties, but due to its linear structure does not provide cryptographic security. Therefore, a stream cipher design based on LFSRs uses a number of different LFSRs and nonlinear Boolean functions coupled in different ways. Three common LFSR-based types of stream cipher can be identified:

- *Nonlinear combination generators.* The keystream is generated as a nonlinear function of the outputs of multiple LFSRs.
- *Nonlinear filter generators.* The keystream is generated as a nonlinear function of stages of a single LFSR.
- *Clock-controlled generators.* In these constructions, the necessary nonlinearity is created by irregular clocking of the LFSRs (e.g., the output from an LFSR is used to clock a second LFSR). The GSM encryption algorithm A5/1 is often described as a stream cipher of this type composed of three LFSRs. One bit of each register determines the clocking. At each step, the registers whose clocking bits share the same value with the clocking bit of some other register, are clocked. Then (at most) one register is in a minority and is not clocked.

The second design strategy for stream ciphers is to iterate some complex function, such as an encryption transformation of a block cipher. Such stream ciphers are defined as modes of operation of block ciphers (see Section 4.2.1.5). The stream cipher designed for the use in UMTS networks falls into this category.

4.2.3 Message authentication codes

4.2.3.1 Security model

A MAC is a cryptographic algorithm that is used to protect the integrity and origin of data. It takes an input of arbitrary length and produces an output of fixed length. A MAC algorithm is a family of functions parameterized using a secret key. Contrary to block ciphers, MAC transformations need not be invertible. The security requirement for a MAC is that, without knowledge of the secret key, it should be infeasible to produce a MAC for any new message, even when some messages and corresponding MAC values are known. A brief overview of the security requirements and general MAC constructions is given below. The interested reader should consult [84], [87] or [131] for further information on MACs.

The MAC of a message X of arbitrary length is computed as a function $H_K(X)$ of X under the control of a secret key K. The length m of the MAC output $H_K(X)$ is fixed. The security requirement is as follows: it must be infeasible, without knowledge of the secret key, to determine the correct value of $H_K(X)$ with a success probability larger than $1/2^m$. This is the probability of simply guessing the MAC value correctly at random. It should not be possible to increase this probability even if a large number of correct pairs X and $H_K(X)$ is available to the attacker.

Similarly to block ciphers, MAC algorithms operate on relatively large blocks of data. Most MACs are iterated constructions. The core function in the MAC algorithm is a compression function. At each round the compression function takes a new data block and compresses it together with the compression result from the previous rounds. Hence the length of the message to be authenticated determines how many iteration rounds are required to compute the MAC value.

For a given message X, its MAC value H can be verified by anybody in possession of the secret key K and the MAC computation algorithm. The MAC system follows the model of the general cryptographic system depicted in Figure 4.1. Alice and Bob share a MAC algorithm and a secret key. Alice has a message X she wants to send to Bob integrity-protected. Alice computes a MAC value $H = H_K(X)$ of X under the control of her secret key K and sends the pair X and H to Bob. When receiving message X, Bob computes a MAC value $H' = H_K(X)$ that checks whether equality $H = H'$ holds. If the two values are equal, Bob can accept the message as authentic (i.e., originally created and sent by Alice).

MACs are also used widely as cryptographic primitives in entity authentication protocols. Consider the situation in Figure 4.1 and assume that Bob wants to verify that it really is Alice he is communicating with. So, Bob sends Alice a challenge X, which typically contains a randomly generated value, but may also be based on a sequence number or time stamp. The main requirement is that the challenge is fresh, unpredictable and nonrepeating. Without unpredictability we have an even more serious flaw: Carol can predict (some) future X. Now Carol can pretend to be Bob against Alice, send X and get H. With this Carol can later pretend to be Alice against Bob. When Alice receives the challenge X, she computes her response as a MAC of X under the control of her secret key and sends it to Bob. Meanwhile, Bob has computed the expected response. When he receives Alice's response he compares it with the expected response. If these two values are equal, Alice has been correctly authenticated.

Unfortunately, this authentication paradigm has a serious flaw. Assume that Carol has access to some network node that is situated in the middle of the communication channel between Alice and Bob. Simply by relaying the challenges and responses, she can pretend to be the end of the communication channel, where Bob is expecting Alice to be. In this case it is said that Carol is acting as the man in the middle between Alice and Bob. The problem is well understood, at least in this simple situation, and many solutions have been developed. One common solution, also used in GSM and UMTS authentication, is that Alice and Bob in addition to response values compute a cryptographic key value that is used to protect the subsequent communication. Carol is still able to copy and send forward anything sent by Alice, but she is not able to decrypt or modify Alice's messages or create her own messages and send them to Bob in the name of Alice.

4.2.3.2 Classification of attacks

The security of a MAC algorithm is usually analysed by determining its resistance against the following types of attacks:

- *Total break*. An attacker finds the secret key K.
- *Selective forgery*. An attacker is able to determine the valid MAC value $H_K(X)$ for a message X chosen by her without knowledge of the secret key K.
- *Existential forgery*. An attacker can determine a valid MAC value $H_K(X)$ for some message X, which she has not seen previously, without knowledge of the secret key K.
- *Distinguishing algorithm*. An attacker is able to tell whether the attacked MAC function is a randomly chosen function with n-bit output or one of the 2^k permutations specified by the MAC algorithm with key length k.

Again, this list of breaks is hierarchical. Practical attacks require that the forgery is verifiable, which means that the attacker knows with large probability that the forged MAC is valid.

The attacks against MAC algorithms can also be classified depending on the information available to an attacker as follows:

- *Known message attack*. The attacker has knowledge of some messages (plaintext) and the corresponding MAC values.
- *Chosen message attack*. The attacker is able to select messages and obtain the corresponding MAC values.
- *Adaptively chosen message attack*. The attacker is able to choose a number of messages and obtain the corresponding MAC values. When choosing a new message she may exploit the information of previously obtained MAC values.

An attacker may try different strategies to forge MAC values. A straightforward way that cannot be prevented is simply to guess the correct MAC value for a chosen message. The probability of success is $1/2^n$, where n is the length of MAC. Increasing the length of the MAC makes it more difficult to guess the MAC value. However, this forgery cannot be verified by the attacker and is hence impractical if the success probability is small.

Success probability can always be somewhat improved using the following strategy. Instead of guessing the MAC value at random the forger may perform some offline analysis of the distribution of the number of keys that have been used to produce certain MAC values. For a fixed message, the number of keys used to produce a MAC value is usually not uniformly random, but some MAC values are obtained for more keys than on average. The average number of keys used to produce a MAC value of length n is 2^{k-n}, where k is the length of the key. Hence the optimal strategy is to select a message and a MAC value that is produced by using the maximum number of keys, say t, where $t > 2^{k-n}$. Then the probability of success is $t/2^k$, which is larger than $1/2^n$. Such an analysis is feasible in practice for short MAC codes with unconditional security. The parameters of computationally secure MACs are prohibitive for performing such analysis and, therefore, simple guessing remains the best strategy.

Another approach is to try and determine the secret key. Similarly to encryption systems, an attacker may perform an exhaustive key search given a valid pair: a message and its MAC value. It takes approximately k/n valid pairs to determine the key of length k uniquely.

More advanced attacks are based on internal collisions of the MAC algorithm. An internal collision occurs when two different messages are input and the compression function transforms them to the same values at some intermediate round. In fact most of the collisions detected in MAC output are caused by internal collisions. Depending on the type of compression function, an internal collision may sometimes be turned to a key recovery attack, which is faster than an exhaustive search. Such an attack was used in [43] to break the early GSM A3/A8 algorithm COMP 128.

Most contemporary MAC algorithms are based on either an encryption function of a block cipher or a cryptographic hash function. The CBC encryption mode of operation can be turned in to a MAC function by simply taking the last ciphertext block to be the MAC value. The basic CBC-MAC has many variations that strengthen its security against various forgeries. A cryptographic hash function provides the desired data compression functionality, but the secret key must be incorporated in the design by additional construction. The Hashing for MAC (HMAC) algorithm [98] is the most frequently used MAC algorithm of this type.

5

3GPP Algorithm Specification Principles

In summer 1999 the 3GPP security architecture for UMTS reached its final form. The SA3 working group of 3GPP was busy on the specification of the main building blocks of the security architecture. Complete drafts of the security protocols for subscriber and network authentication as well as voice and data encryption were already in place and interfaces to the cryptographic algorithms had been specified. Voice and data need to be protected when transmitted over the radio link between the mobile terminal and the network, and so an encryption algorithm was required to protect the confidentiality of the traffic and an integrity algorithm was required to protect the traffic (especially signalling messages) against malicious modification. However, cryptographic algorithms were not only missing they needed to be standardized as well.

The GSM system also employs cryptographic solutions. A standard algorithm known as A5 is used to encrypt traffic between the mobile terminal and the base station. The A5 algorithm had created a lot of controversial discussion in the press: it had been specified as a secret algorithm, but in the late 1990s it was reverse-engineered by a team of cryptographers at the University of California at Berkeley and found to be of moderate strength only.

One of the goals of 3G had been to fix the problems found in 2G. The publicity surrounding secret 2G cryptographic algorithms was considered problematic, it being understood that the secrecy approach applied for GSM was no longer feasible. The benefits of making cryptographic algorithms public were evident. It would be very difficult to gain public confidence in UMTS security if the cryptographic solutions were kept secret, but it was also realized that making the algorithms public would impose particular requirements on the design process. The common trend was toward using publicly available cryptographic algorithms.

In summer 1999, SA3 had the task of defining and agreeing on the design process for encryption and integrity algorithms. Previous examples of design processes of cryptographic algorithms intended for use in public systems existed, but there was no standard strategy for performing such a task. The first such effort was the design of the public DES by NIST in 1977. In the area of telecommunication, public cryptographic solutions were designed for the Wireless Local Area Network (WLAN) standard IEEE 802.11 (published in 1997) and Bluetooth (published in 1999) (see also Section 3.3.2).

In 1999 NIST had just started the AES project with the remit of finding a replacement for the DES. However, the results of the NIST project would not be available until 2001 and, as a 128-bit block cipher, the AES would probably be too large to fit within the 10,000 gates of hardware that had been specified as the maximum size for handset implementation.

Facing the task of selecting confidentiality and integrity algorithms, in summer 1999 SA3 made an assessment of the different options it had for performing this task. The results of this study are documented in [15] and were accepted by SA3 in July 1999. This study focused on the design of confidentiality and integrity functions that were finalized in Release 1999. Later, the document was included in Release 4 without any technical changes. In [15] the following three specification strategies were identified:

1. select an off-the-shelf algorithm;
2. invite submissions;
3. commission a special group to design an algorithm.

It was realized that different strategies have different implications for suitability, security and timely delivery of the algorithm. However, the feasibility of each strategy is based on different assumptions about the availability of such resources as expert knowledge and time.

Whichever of the three specification strategies is selected, it was understood that a separate strategy must be defined for security evaluation of the specified algorithm before it is adopted for use. The evaluation can either rely on voluntary efforts, or special groups of experts could be commissioned. For open designs, voluntary efforts will become available as soon as the algorithm is published. During the 1990s, the formerly secret art of cryptanalysis had developed into an open science and researchers were continuously looking for suitable objects for study. Because security flaws had to be identified before publication of the algorithm and because only limited time was available, it was not an option to rely on voluntary efforts.

One of the most important conclusions of [15] was that if the algorithm is open to the public, then the analysis methods and results should be published together with the evaluation report. This is important for achieving the necessary confidence in the algorithm. It was also understood that if the evaluation was carried out on a secret algorithm or if just the conclusions were published, then trust in the algorithm would to a large extent depend on trust in the experts.

The principles defined in [15] determined to a large extent the design and evaluation strategies that were adopted for the design of the standard 3GPP confidentiality and integrity algorithms for UMTS. Later the design effort for the Authentication and Key Agreement (AKA) functions were based on the same strategies. Recently, the TIPHON group of the European Telecommunications Institute (ETSI) included them as part of the common ETSI guidelines on evaluation criteria for cryptographic algorithms [51].

Confidentiality and integrity algorithms were to be delivered by December 1999, so only limited time was available. The adopted approach was to commission one group of experts to design the algorithms and one or more groups of experts to analyse them. The algorithms should in the end be made public (after expert evaluation was finished) to achieve maximum public confidence in the systems. A suitable group to be commissioned for the task was near to hand: for about eight years a group called ETSI SAGE had been designing and evaluating cryptographic algorithms for ETSI (an overview of the work of SAGE is given in [125]). This group was nominated the design authority for the algorithms. In addition SAGE was instructed to strengthen the group by drawing on appropriate expertise within the 3GPP organizations in addition to its normal resource pool of experts. Four 3GPP companies—Ericsson, Mitsubishi, Motorola and Nokia—volunteered to provide one expert each for this work.

In [15] several liability issues were also studied.

In late summer 1999 a special task force was set up to design the standard confidentiality and integrity algorithms for UMTS. The design principles, specification and evaluation results of this task force are described in Chapter 6. In 2000 the task force was called on again, this time to design an example algorithm for UMTS AKA functions. AKA algorithms need not be standardized and each operator can use its own algorithm for authorization of its subscribers' access to the UMTS network. The culmination of this work was the MILENAGE parameter: its requirements, specification and evaluation are described in Chapter 8. The open cryptanalytic research community has studied 3GPP algorithms, and a number of research papers have already been published in workshops and conferences. Let us now look at the main part of these results.

6

Confidentiality and Integrity Algorithms

Confidentiality and integrity algorithms were constructed together as special, new block cipher modes (see Section 4.2.1.5). The block cipher itself is also a new design and is described in Chapter 7. First, the requirements for UMTS integrity and confidentiality algorithms will be reviewed in Sections 6.1 and 6.2. Then, the specification process as well as the initial choices and design principles are described in the subsequent four sections, where the conclusions of the evaluation by the task force are also given, to give the reader an idea of the methods used by the task force to assess the algorithms at the time of delivery. Details of the evaluation and other cryptanalysis are described in Sections 6.6.3 and 6.8.3 and the specification of the 3GPP f8 stream cipher mode is given in Section 6.6. In Release 6 the original f8 mode has also been extended to provide new, strong encryption algorithms for GSM, ECSD (Enhanced Circuit Switched Data) and GPRS (General Packet Radio Service). These will be described in Section 6.7 and the integrity mode of operation is given in Section 6.8.

6.1 Requirements for the Confidentiality Algorithm

6.1.1 Functional requirements

The requirements for UMTS confidentiality and integrity algorithms are specified by 3GPP in the technical specification document 33.105. The latest (Release 4) specification is given in [10].

The mechanism for data confidentiality of user data and signalling data in UMTS requires a cryptographic function called f8. As specified in [10 sect. 5.2.6], function f8 must be a synchronous stream cipher algorithm (see Section 4.2.2.1) and for interoperability within UMTS it must be fully standardized. It will only be used to protect the confidentiality of user data and signalling data sent over the radio access link between the User Equipment (UE) and the Radio Network Controller (RNC) and, therefore, is implemented in the UE and the RNC. Encryption will be applied in the Medium Access Control (MAC) sublayer and in the Radio Link Control (RLC) sublayer of the data link layer (layer 2).

In the UE the algorithm may be implemented as hardware, while in the RNC it may also be implemented in software on a general-purpose processor. Therefore, the algorithm should be designed to accommodate a range of implementation options. For hardware implementations, the working assumption was that it should be possible to implement one instance of the algorithm using less than 10,000 gates.

A wide range of UE with different bearer capabilities is expected, so encryption throughput requirements on the algorithm will vary depending on the implementation. However, based on likely maximum user traffic data rates, it must be possible to implement the algorithm to achieve an encryption speed in the order of 2 Mbit/s at downlink and uplink. Encryption throughput requirements should be met based on clock speeds of 20 MHz.

The algorithm is used in three modes and nominates a different type of data for each mode. The requirements for the keystream generator in each mode are as follows:

1. In RLC-transparent mode a new keystream block of 10 ms is required for each new, physical layer frame. The length of a frame varies from 1 bit to 20,000 bits, with granularity of 1 bit.

2. In UM (Unacknowledged Mode) RLC mode a new keystream block is required for each new PDU (Protocol Data Unit). The length of PDU varies from 16 bits to 5,000 bits, with granularity of 8 bits.

3. In AM (Acknowledged Mode) RLC mode a new keystream block is required for each new PDU. The length of PDU varies from 24 bits to 5,000 bits with granularity of 8 bits.

6.1.2 Algorithm operation

The f8 function is a synchronous stream cipher whose encryption and decryption operations are based on the same secret key and the same set of values as the initialization parameters.

Encryption and decryption operations for a synchronous stream cipher are the same. For both operations a keystream block is generated. The values of the keystream bits depend on the given key and initialization parameters. The keystream block and the received data block are added together using a bitwise XOR operation. The length of the data block determines the length of the keystream block.

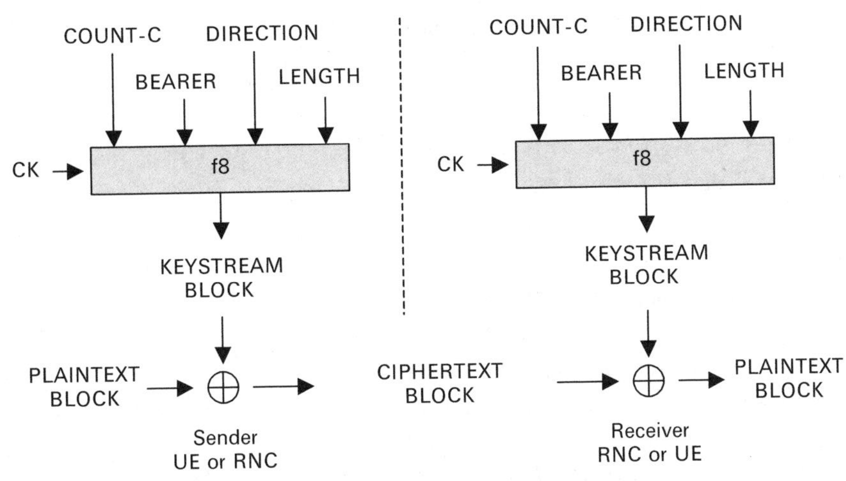

Figure 6.1 Ciphering user and signalling data transmitted over the radio access link
CK = Cipher Key; UE = User Equipment; RNC = Radio Network Controller

Figure 6.1 illustrates the use of f8 to encrypt plaintext by applying a keystream using a bitwise XOR operation. The plaintext may be recovered by generating the same keystream using the same input parameters and applying it to ciphertext using a bitwise XOR operation.

The input parameters to f8 are the Cipher Key (CK), the time-dependent input (COUNT-C), the bearer identity (BEARER), the direction of the transmission (DIRECTION) and the length (LENGTH) of the plaintext block. The format of the parameters is defined in the next section.

The CK is renewed at every authentication process. COUNT-C, BEARER and DIRECTION can be considered as initialization parameters as they are renewed for each keystream block. The time-dependent input COUNT-C is also sent in cleartext and used as a synchronization parameter for the synchronous stream cipher. The input parameter LENGTH only affects the length of the keystream, not the actual bits in it.

Based on these input parameters the algorithm generates the output keystream block (KEYSTREAM), which is used to encrypt the input plaintext block (PLAINTEXT) and to produce the output ciphertext block (CIPHERTEXT).

6.1.3 Interfaces to the algorithm

6.1.3.1 Cipher Key (CK)

The length of CK is 128 bits. In case the length k of the generated key is smaller than 128 bits, the most significant bits of CK carry the nominal key information, whereas the remaining bits repeat the key information as follows:

$$CK = CK[0], CK[1], \ldots, CK[127]$$

where $CK[0], \ldots, CK[k-1]$ carry the key information and $CK[n] = CK[n \bmod k]$, for all n, such that $k \leq n < 128$.

6.1.3.2 Time-dependent input (COUNT-C)

The length of the COUNT-C parameter is 32 bits. Synchronization of the keystream is based on the use of a physical layer frame counter combined with a hyperframe counter introduced to avoid reuse of the keystream. The counter (COUNT-C) is initialized at connection establishment. The exact structure of COUNT-C is specified in TS 33.102 [1] (see also Section 2.1.3).

6.1.3.3 Radio bearer identity (BEARER)

The length of BEARER is 5 bits. The same cipher key may be simultaneously used for different radio bearers associated with a single user. To avoid using the same keystream to encrypt more than one bearer, the algorithm generates the keystream based on the identity of the radio bearer.

6.1.3.4 Transmission direction (DIRECTION)

The value of the DIRECTION bit is 0 for uplink (messages from UE to RNC) and 1 for downlink (from RNC to UE). The same cipher key may be used for uplink and downlink channels. The purpose of the DIRECTION bit is to avoid using the same keystream to encrypt both uplink and downlink transmissions.

In the GSM separate segments of the output of the A5 algorithm are used for encryption of the uplink and downlink traffic, while in the UMTS an explicit direction value is required to perform the separation.

6.1.3.5 Keystream length (LENGTH)

LENGTH is an integer between 1 and 20,000. The length of LENGTH is 16 bits. For a given bearer and transmission direction the length of the plaintext block that is transmitted during a single physical layer frame may vary. The algorithm will generate a keystream block of variable length based on the value of the LENGTH parameter.

The range of values of the LENGTH parameter will depend not only on the RLC PDU/MAC SDU (Signalling Data Unit) size but also the number of RLC PDUs/MAC SDUs that may be sent in a single, physical layer, 10-ms frame for a given bearer and transmission direction. Further, it is specified in [10]:

> Not all values between the maximum and minimum values shall be required but it is expected that the ability to produce length values of whole numbers of octets between a minimum and a maximum value will be required.

Earlier in the same document (see also Section 6.1.1) it was mentioned that granularity of the physical layer frame length is 1 bit. The standard f8 algorithm is specified to support any lengths and length granularities.

6.1.3.6 KEYSTREAM

The length of a keystream block equals the value of the input parameter LENGTH.

6.1.3.7 PLAINTEXT

The length of a plaintext block equals the value of the input parameter LENGTH. This plaintext block consists of the payload of the particular RLC PDUs/MAC SDUs to be encrypted for a given bearer and transmission direction.

6.1.3.8 CIPHERTEXT

The length of a ciphertext block equals the value of the input parameter LENGTH.

6.2 Requirements for the Integrity Algorithm

6.2.1 Overview

A cryptographic function (f9) is used to protect data integrity and authenticate the data origin of signalling data at the RRC layer. The f9 function is implemented in the UE and the RNC, and to support interoperability it must be fully standardized. Similarly to the confidentiality algorithm, the integrity algorithm should be designed to accommodate a range of implementation options including hardware and software implementations.

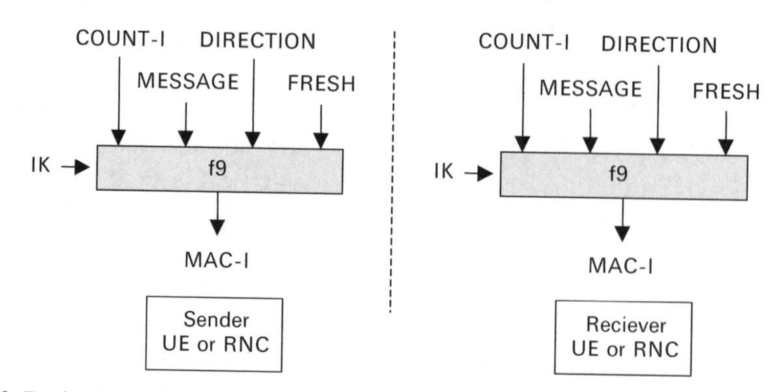

Figure 6.2 Derivation of MAC-I (or XMAC-I) on a signalling message
MAC-I = Message Authentication Code; XMAC = expected MAC; IK = Integrity Key; UE = User Equipment; RNC = Radio Network Controller

A cryptographic message authentication algorithm generates a fixed length MAC from a message of arbitrary length, under the control of the secret parameter key and a set of initialization values. The sender and the receiver generate the MAC using the same function. The sender sends its MAC result to the receiver, who compares the received MAC value with the expected MAC value computed by the receiver. The receiver accepts the MAC if the compared values are equal.

Figure 6.2 illustrates how f9 is used to derive a MAC-I (I = integrity of signalling data) on a signalling message.

The input parameters to the integrity algorithm are the Integrity Key (IK), a time- and frame-dependent input (COUNT-I), a random value generated by the network side (FRESH), the direction bit (DIRECTION) and the signalling message (MESSAGE). The exact format of these parameters is given in Section 6.2.2.

IK is a cryptographic key that is newly generated at each authentication process. The COUNT-I, FRESH and DIRECTION parameters can be considered as a set of initialization parameters that are renewed for each message to be authenticated.

Based on these input parameters the user computes using the f9 function the MAC-I for data integrity, which is appended to the message when sent over the radio access link. The receiver computes the expected MAC value (XMAC-I) on the messages received in the same way as the sender computed MAC-I on the message sent.

6.2.2 Interface

6.2.2.1 Integrity key (IK)

The length of IK is 128 bits.

6.2.2.2 Frame-dependent input (COUNT-I)

The length of COUNT-I is 32 bits. It protects against replay during a connection because its value is incremented by 1 for each input message. COUNT-I consists of two parts: the Hyper Frame Number (HFN) as the most significant part and a Radio Resource Control (RRC) sequence number (SQN) as the least significant part.

The COUNT-I parameter is initialized at connection set-up using the same procedure as for COUNT-C. The initial value of the HFN is sent by the user to the network at connection set-up. The user stores the most significant part of the greatest-used HFN from the previous connection and increments it for the new connection. In this way the user is assured that no COUNT-I value is reused (by the network) using the same IK. The exact structure of COUNT-I is specified in TS 33.102 [1].

6.2.2.3 One-time random number (FRESH)

The length of FRESH is 32 bits. The same IK may be used for several consecutive connections. This FRESH value is input to the algorithm to assure the network side that the user is not replaying old MAC-Is.

6.2.2.4 The signalling data (MESSAGE)

Document 33.105 does not explicitly specify the maximum length of the signalling data that are input to the f9 function. In December 2001 at 3GPP TSG SA plenary #14 two change requests were approved for TS 35.201: the first was to increase the maximum length of f8 and f9 input data to 20,000 bits in Release 1999 and the second was the same for Release 4. The reason for requesting the changes was that previously the maximum length of the keystream was changed from 5,114 to 20,000 in TS 33.105. The change requests aimed to bring inconsistent specifications into alignment and, more importantly, prevent implementations from taking 5,114 bits as an upper limit. Though it is mentioned in the change request document that the limit of 20,000 will most likely never be reached for f9 data input.

The changes were incorporated in new versions 3.2.0 (Release 1999) and 4.1.0 (Release 4) of TS 35.201. The later-published version 5.0.0 (Release 5) is identical. In the changed documents the maximum value of the keystream length generated by f8 is defined to be 20,000, while no limit is put on the input message length of f9.

6.2.2.5 Direction of transmission (DIRECTION)

The length of DIRECTION is 1 bit. The same IK may be simultaneously used for uplink and downlink channels associated with the UE. The value of the

DIRECTION is 0 for messages from the UE to the RNC and 1 for messages from the RNC to the UE.

6.2.2.6 Message Authentication Code (MAC-I) and expected MAC-I (XMAC-I)

The length of MAC-I is 32 bits. The UMTS security specification makes use of two other MACs—MAC-A and MAC-S—which are used in the authentication process, (see [10] and Section 8.2.1).

6.3 Design Task Force

The task of designing and formulating the specification of standard confidentiality and integrity algorithms was given to SAGE. Under the leadership of SAGE, ETSI set up a specific project, called the SAGE Task Force for 3GPP algorithms (SAGE TF 3GPP) for this purpose. A detailed report of the algorithm specification project was given by the Task Force itself in [16], which the following description draws on. Another account of the work of the algorithm Task Force was given by Steve Babbage in [36].

The Task Force comprised the regular SAGE members, the designer of the MISTY algorithms and three manufacturers from the 3GPP. The work was funded by 3GPP and the three manufacturers. SAGE TF 3GPP work was thus carried out by 11 organizations, which were divided into two teams: a design team and an evaluation team.

The design of the algorithms and a complete set of specification documents were finalized in mid-November 1999. It was decided by 3GPP that the algorithms should be evaluated over a period of one month by three groups of independent evaluators, all consisting of well-known cryptologists. The three groups of independent evaluators were a consortium led by Leuven University (Belgium), Cryptolog in Paris and the Royal Holloway College, University of London. The results of these evaluations were reviewed by the Task Force before final algorithm specifications were released to the 3GPP.

6.4 Getting Started

In summer 1999 when SAGE took on the design task for standard 3GPP confidentiality and integrity algorithms for the UMTS, the most important document at their disposal was the newly-published requirement specification TS33.105 [10] (version 3.0.0). This document defines the security services that cryptographic algorithms should provide for security of the UMTS. It also defines the technical interfaces, such as input and output parameters, as well as performance requirements. These requirements were reviewed in Sections 6.1 and 6.2.

The first task was to define a starting point for the work, review the requirement specification and then derive relevant cryptographic criteria from the general security requirements for the UMTS. The results of this work as reported in [16] are given in Section 6.4.3.

6.4.1 SAGE contribution to SA3

During the initial study period SAGE also noticed that the security services provided by the integrity algorithm needed to be improved. In August 1999 a liaison statement from SAGE was presented to the SA3 subgroup related to this issue. SAGE observed that it was necessary to strengthen the integrity algorithm in two ways: uplink and downlink needed to be cryptographically separated and the length of MAC output was too low according to current standards. SAGE proposed two enhancements to the UMTS security specification:

1. increase the bit-length of MAC output from 24 to 32;
2. add a bit that indicates the direction of inputs.

These proposals were subsequently included in version 3.1.0 of [10]. MAC length 32 was selected as a compromise between security and the limited bandwidth resources over the air interface. For more discussion on MAC length, see Section 6.8.3.

6.4.2 Modes around MISTY1

SAGE made the decision to adopt the MISTY1 block cipher algorithm as the starting point for the 3GPP algorithm work at an early stage. The plan was to build the algorithms as special modes of operation around a possibly modified version of MISTY1. An alternative approach for a stream cipher design would have been a dedicated stream cipher that does not rely on a previously existing cryptographic algorithm. There were two major reasons in favour of a design using a block cipher. First, getting part of the design off the shelf would speed up the design and evaluation procedure and, second, the integrity algorithms are typically MAC algorithms constructed on hash functions or block ciphers (Section 4.2.3).

Given the short timescales and the requirement of an open design, SAGE decided to apply the off-the-shelf approach and see what was readily available that could be reused for that purpose. Three years earlier, in 1996, Mitsuru Matsui from Mitsubishi Electrical Corporation had proposed two constructions for a 64-bit block cipher, which were provably secure against linear and differential cryptanalysis [82]. The proofs were based on the theory developed by Kaisa Nyberg and Lars Knudsen [94], [95]. The following year Matsui published the completed designs for 64-bit block ciphers in [83], which he named MISTY1 and MISTY2. Matsui's algorithms attracted SAGE for at least the following reasons:

1. they were designed to be suitable for implementation in software or hardware;
2. they had already been scrutinized by voluntary research efforts to some extent;
3. they had a quantitative basis for their security;
4. the length of the key was 128 bits.

Moreover, Mitsuru Matsui was able to join the algorithm development Task Force and helped make further modifications and adaptations to MISTY algorithms. Early cryptanalysis of the MISTY ciphers had revealed that MISTY1 had some advantages over MISTY2 with respect to pseudorandomness properties (see also Section 7.2.2). This may have been a further reason that SAGE selected MISTY1.

The number of high-quality block ciphers with 64-bit block size and 128-bit key available off-the-shelf in 1999 was not large. No more than four other candidates could seriously be considered as alternatives to MISTY: Triple-DES, IDEA, SAFER K-128 and RC5. Out of these five ciphers, MISTY1 was considered to be best suited as the "crypto engine" for 3GPP confidentiality and encryption algorithms.

The starting point for the 3GPP confidentiality algorithm was the standard OFB mode and for the integrity algorithm the CBC-MAC. To test the feasibility of this approach, particularly from the point of view of performance in hardware, SAGE distributed initial draft designs, with MISTY1 as the block cipher, to a number of 3GPP member organizations to review its performance and implementation complexity characteristics. This resulted in positive responses and no indications that the algorithm would be too complex in implementation or slow in operation. According to initial estimates it would be possible to implement MISTY1 in less than 3,000 gates. The upper bound to the total number of gates for the f8 (and f9) function was set at 10,000 in [10].

6.4.3 Particular security criteria

The specific use and area of application intended for f8 and f9 algorithms also placed additional requirements on the algorithms that had to be taken into consideration. In [16, sect. 8.1.4] these particular conditions are reported as follows.

6.4.3.1 MISTY

The most successful attacks on MISTY are higher-order differential attacks and interpolation attacks, and any modifications to MISTY must not increase the algorithm's susceptibility to these attacks—ideally, they should strengthen it against them.

6.4.3.2 General

These algorithms use short-term keys that are presented by the USIM or SIM (Subscriber Identity Module) to the phone. There is therefore no requirement for resistance to differential power analysis (aimed at extracting a long-term key from supposedly secure storage). Keys are always randomly generated, so related key attacks are of very little practical significance.

6.4.3.3 Confidentiality algorithm f8

The most popular way of building a stream cipher from a block cipher is to use it in OFB mode (see also Section 4.2.2.3). This approach carries a (very small) risk of short keystream cycles, but it would be desirable to remove this risk altogether. It is also desirable to avoid any possibility of the keystream generator getting itself into exactly the same state—and hence generating the same keystream from then onward—at any two points in different frames.

6.4.3.4 Integrity algorithm f9

For absolute security, it must be impossible for an attacker, after intercepting one message and a MAC pair, to modify the message in any way and have the MAC either unchanged or modified in a way she can predict (e.g., linearly). In particular, therefore, the padding applied to a message to bring it to a whole number of blocks must be such that it is not feasible to construct two messages that are identical after padding.

6.5 Design Process

6.5.1 The teams

The SAGE 3GPP TF operated in two teams: the design team and the evaluation team. The design team was responsible for proposing design and drafting design specifications, while the evaluation team was responsible for cryptanalysis of the proposed designs as well as their statistical evaluation.

As reported in [16]: "The algorithms were designed using the iterative, interactive and phased approach that is normally applied for the design of ETSI SAGE algorithms." The final design was reached in four phases, during which the algorithms were analysed, alternative designs proposed and discussed, and decisions taken. The phases of the design process are outlined in [16].

By mid-November 1999 the KASUMI, f8 and f9 algorithms were fully fixed and a final round of statistical tests on the algorithms was carried out. The specification documents were drafted and two parties independently carried out specification testing to check the correctness and completeness of the specification.

The specification document and statistical test data as well as a summary of the evaluation undertaken by the SAGE TF 3GPP were then made available to three groups of independent evaluators. Over four weeks the algorithms were evaluated by three groups of independent evaluators. This resulted in three evaluation reports and the reports were reviewed by the SAGE TF 3GPP. After this review the algorithm specifications were finalized.

6.5.2 Design documentation

TS 33.105 [10] also sets the requirements for specification documentation, which in the case of the 3GPP confidentiality and integrity algorithms comprises four documents. In addition, a design and evaluation report is required by [10] for the purposes of quality assurance. The Task Force produced two design and evaluation reports: one to describe the conducted work and one to give a detailed account of the results of the security evaluation process.

6.5.2.1 Technical specification documents

The specification for the confidentiality and integrity algorithms is split into two documents. This is done for purely practical reasons and does not indicate that the specifications or the described algorithms could be applied independently. Document 1 [19] contains the f8 and f9 specifications and document 2 [20] is the KASUMI specification.

An unambiguous algorithm specification is intended for use by implementors of the algorithms. The specification includes annexes that provide simulation code for the algorithms written in ANSI (American National Standards Institute) C. The specification also includes annexes that contain illustrations of the functional elements of the algorithms.

Document 3 [21] of the specification set contains implementor test data, provided to assist implementors of the algorithm to realize the algorithm specification. This set of test data, as well as including algorithm input and output data, includes details of the internal state of the algorithm at various stages in its execution. Sufficient detail has been provided to enable implementors to readily identify the likely location of any errors in their implementation.

Document 4 [22] contains the design conformance test data, produced to allow implementors of the algorithm to validate their implementations and manufacturers to validate hardware embodiments of the algorithm (e.g., in ASICs (Application Specific Integrated Circuits) or FPGAs (Field-Programmable Gate Arrays)). The test data set is presented as input/output test data, allowing realization to be tested as a "black box". The design conformance test data are designed to give a high degree of confidence in the correctness of any implementation of the algorithm. The set of test data ensures that all elements of the algorithm are fully exercised.

6.5.2.2 Technical reports

In addition to the specification document any algorithm project must also produce a design and evaluation report, whose purpose, as defined in [10], is to provide evidence to potential users of the algorithm, specification and test data that appropriate and adequate quality control has been applied to their production. The report must explain the following:

- the algorithm and test data design criteria;
- the algorithm evaluation criteria;
- the methodology used to design and evaluate the algorithm;
- the extent of mathematical analysis and statistical testing applied to the algorithm;
- the principal conclusions made about evaluation of the algorithm;
- the quality control applied to the production of the algorithm specification and test data.

The report must confirm that all members of the design authority have approved the algorithm, specification and test data and contain key conclusions from a commissioned, closed evaluation of the algorithm.

SAGE TF 3GPP decided to prepare two technical documents to cover the material required for a design and evaluation report. The first is entitled *Specification and Evaluation of 3GPP Standard Confidentiality and Integrity Algorithms* and covers the requirements listed above.

The title of the second technical document is *Report on the Evaluation of 3GPP Standard Confidentiality and Integrity Algorithms*. Such a report was not required to be included in the standard documentation and the main reason for producing it was the intended publication of the algorithms. The Task Force considered that published algorithms should be accompanied by a report that provides detailed information about the design principles and the security evaluation of the algorithms. It was felt that unjustified claims and misinterpretations about the security of the algorithms could be prevented if the design principles and security level were known.

However, publication of the evaluation report was delayed. After the algorithms were published, the evaluation document was available at a public 3GPP document site for a few weeks. It then took more than a year before it was published as an official 3GPP document, as TR33.909[1] [17]. Meanwhile, the published f8, f9 and KASUMI algorithms were subjected to external cryptanalysis and as a result one research paper appeared that contained cryptanalysis of KASUMI already known to the Task Force (for more details, see Section 7.6).

6.5.3 Conclusion of evaluation

Time is considered as one of the most important resources in the evaluation of security systems, in general, and cryptographic algorithms, in particular. In 3GPP algorithm specification work the available time was very limited. Shortage of time can be partly compensated for by involving more experts, but not completely. It is worth noting that three years later none of the conclusions turned out to be false or even suspect. The conclusion written by the Task Force in December 1999 and published in the evaluation report [17] is reproduced below in full:

> The 3GPP confidentiality and integrity algorithms have been subject to an extensive mathematical and statistical review in order to reveal any weakness in the design. This work has been conducted by the Task Force itself, by additional manufacturers with competence in the field, and by three independent parties. The work has involved some of the leading experts in the field. *The general conclusion is that the algorithms are based on sound design principles, and no practical attacks were found. The algorithms are well fitted for their intended use* [Task Force's italics].
>
> The algorithms have specifically been designed for use within the 3GPP context. It has not been the intention to increase the security margins in

[1] Currently, only version 1.0.0 exists. Although document TR33.909 exists in Release 4, versions 4.0.0 and 4.1.0 contain a different document.

order to develop general-purpose algorithms for multiple unknown applications. The design is a careful trade-off providing full strength algorithms and efficient implementation and use in the next generation mobile systems.

The 3GPP algorithms have been designed to resist a suite of well-known cryptanalytic attacks. However, one can never prove that a cryptographic algorithm will resist new attacks in the future. Due to this fact and the very limited time span that was available for the work, the Task Force will propose that the results from this report are reviewed on a regular basis. A basic review of the offered security and usability of the 3GPP confidentiality and integrity algorithms should be conducted every five years.

6.6 Confidentiality Algorithm

6.6.1 The f8 stream cipher mode

Confidentiality algorithm f8 is a stream cipher that is used to encrypt and decrypt blocks of data under a confidentiality key (CK). The block of data can be between 1 and 20,000 bits long. The algorithm uses KASUMI in a form of OFB mode as a keystream generator. The f8 algorithm is specified in TS 35.201 [19].

The 3GPP f8 stream cipher mode is not a standard stream cipher mode of operation of a block cipher. Examples of such standard modes are counter mode and OFB mode (see Section 4.2.1.5 or [84]). A counter mode keystream generator makes use of a counter that is updated for each new block and is taken as part of the input to the generator function. The f8 stream cipher mode can be seen as a combination of these two standard modes and makes use of prewhitening of feedback data. These three features—output feedback, counter and prewhitening—are combined in the following manner. Before the newly generated keystream block is taken back as input to the generator function it is modified by the counter value and the prewhitening data block, using a bitwise XOR operation, which is depicted in Figure 6.3.

6.6.2 Description of f8

The f8 algorithm makes use of the KASUMI key-dependent function, which operates on 64-bit data blocks and produces 64-bit blocks under control of a 128-bit key K.

Inputs to the f8 algorithm are as defined in Section 6.1.3. The algorithm makes use of two 64-bit registers: the static register A and the counter BLKCNT. Register A is initialized using the 64-bit initialization value:

$$IV = \text{COUNT} \,\|\, \text{BEARER} \,\|\, \text{DIRECTION} \,\|\, 0 \ldots 0$$

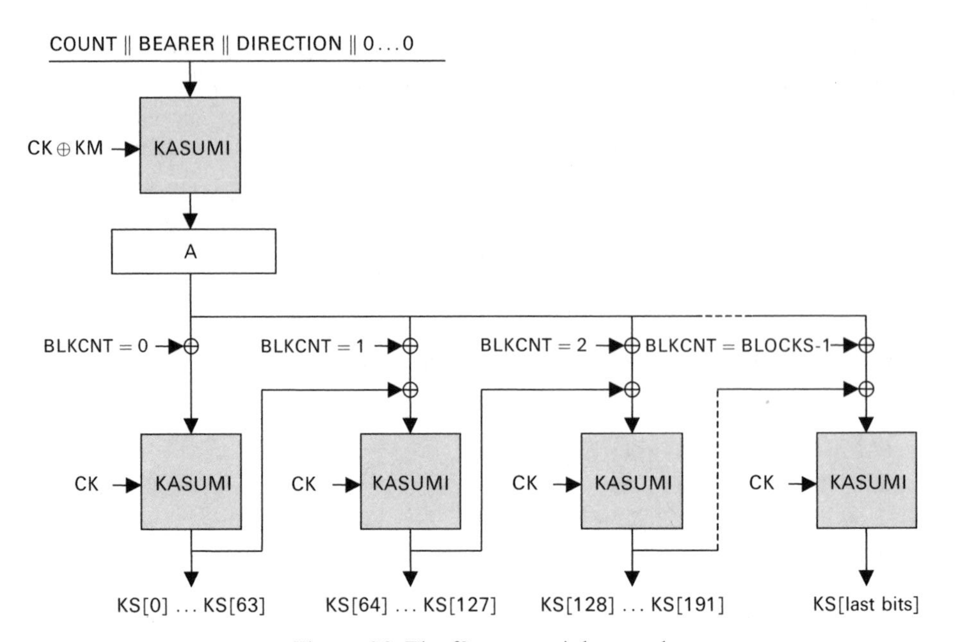

Figure 6.3 The f8 stream cipher mode
CK = Confidentiality Key; KM = Key Modifier; BLKCNT = block counter; KS = keystream

obtained as the concatenation of the 32-bit COUNT, 5-bit BEARER, 1-bit DIRECTION values and a string of 26 zero bits. The counter BLKCNT is set to 0.

The f8 algorithm makes use of a Key Modifier (KM) constant that is equal to the octet $0 \times 55 = 01010101$ repeated 16 times. First, a single operation of KASUMI is applied to register A, using a modified version of the CK to compute the pre-whitening value:

$$W = \text{KASUMI}_{\text{CK} \oplus \text{KM}}(IV)$$

which is stored in register A. Once the keystream generator has been initialized in this manner, it is ready to be used to generate keystream bits. The plaintext/ciphertext to be encrypted/decrypted consists of LENGTH bits, where LENGTH varies between 1 to 20,000 with granularity of 1 bit, while the keystream generator produces keystream bits in multiples of 64 bits. Between 0 and 63 of the least significant bits are discarded from the last block depending on the total number of bits required by LENGTH.

The number of required keystream bits is denoted by BLOCKS, whose value is determined by the value of the LENGTH parameter as follows: the value of LENGTH is divided by 64 and the result is rounded up to the nearest integer. The keystream blocks are denoted as $\text{KSB}_1, \text{KSB}_2, \ldots, \text{KSB}_{\text{BLOCKS}}$. Set $\text{KSB}_0 = 0$

and let n be an integer with $1 \leq n \leq \text{BLOCKS}$, such that $n = \text{BLKCNT} + 1$, and set:

$$\text{KSB}_n = \text{KASUMI}_{\text{CK}}(W \oplus (n-1) \oplus \text{KSB}_{n-1})$$

Individual bits KS[0], KS[1], ..., KS[LENGTH − 1] of the keystream are extracted in turn from KSB_1 to $\text{KSB}_{\text{BLOCKS}}$, with the most significant bit extracted first, by applying the following operation. For $n = 1, \ldots, \text{BLOCKS}$ and for each integer i, with $0 \leq i \leq 63$, set:

$$\text{KS}[((n-1) * 64) + i] = \text{KSB}_n[i]$$

Encryption/decryption operations are identical and are carried out by the bitwise XOR of the input data using the generated keystream.

6.6.3 Security

6.6.3.1 Security claims by the Task Force

In the public evaluation report TR 33.909 [17] arguments are given in support of the f8 construction. The f8 construction uses a prewhitening field $W = \text{KASUMI}_{\text{CK} \oplus \text{KM}}(IV)$, which is considered to have two advantages. First, it offers protection against chosen plaintext attacks. If the initial value $IV = \text{COUNT} \parallel \text{BEARER} \parallel \text{DIRECTION} \parallel 0 \ldots 0$ is used directly in the OFB chain, then, for a known plaintext, the input and output of each KASUMI operation are known, while with prewhitening the plaintext inputs to the KASUMI are no longer known. It also offers protection against collision attacks. Analysis shows that it would, at least in principle, be possible to distinguish the pseudorandom function generator associated with an f8 construction without prewhitening from a truly random generator, based on the observation of 2^{33} keystream blocks. However, this does not seem to be the case with the actual f8 construction for the following two reasons:

- For distinct initial values IV and IV', the prewhitening constants W and W' differ, making it difficult to predict whether keystream blocks associated with IV and IV' are equal. However, the observation of two equal keystream blocks associated with two distinct IV values provides an adversary with the value $W \oplus W'$, but this does not of itself represent a "distinguishing event", though it allows us to predict which other pairs of blocks associated with IV and IV' are equal. But since there are less than 80 blocks in both keystreams, the probability of such a distinguishing event is less than $80^2/2 \cdot 2^{64} \approx 2^{-52}$.

- Given a fixed initial value IV, collisions on keystream blocks are predictable, but the probability that such collisions occur among the at most 80 blocks of the keystream associated with IV is less than $80^2/2 \cdot 2^{64} \approx 2^{-52}$. So even if an adversary is provided with the keystream sequences associated with all possible count values, distinguishing probability remains low.

These estimates of probabilities of undesired events are based on the previously-given upper boundary 5,114 of the bit-length of the plaintext that is to be encrypted using the same initialization of the stream cipher. After the maximum length was increased to 20,000 bits, these derived probabilities need to be multiplied by a factor of $2^4 = 16$, resulting in no significant change to the conclusions derived from this security analysis.

6.6.3.2 Toward a security proof

The security of an encryption scheme is often evaluated in the security model of left or right distinguishability, developed by Mihir Bellare et al. in [39]. This notion of security seems to capture the essential security requirements and allows security proofs that are genuine. In this model, a symmetric encryption scheme is said to be secure if it resists the following distinguishing attack. The attacker is allowed to choose a number of ordered pairs of plaintexts, a left plaintext and a right plaintext, and give them to the encrypting oracle, which implements the encrypting transformation with a randomly chosen key. When receiving left and right plaintexts, the oracle selects one of them randomly, encrypts it and gives it back to the attacker. The attacker wins if she can tell which of the two plaintexts was encrypted by the oracle.

In their paper [39], Bellare et al. proved the security of a block cipher's three stream cipher modes of operation. These were the counter mode, the XOR mode (a variation of the counter mode) and the standard CBC mode. Later, in a paper presented at *Asiacrypt 2001* [67], Ju-Sung Kang et al. used the same security model for the 3GPP f8 construction, but the work is still in progress.

6.7 Extension of the UMTS Confidentiality Algorithm

6.7.1 Background

In spring 2002 SAGE was given the task to design a new encryption algorithm for GSM, ECSD, GPRS and EGPRS (enhanced GPRS) encryption. The new algorithm was intended to be implemented in dual-mode handsets (i.e., handsets operating in both GSM and UMTS modes). A natural design choice was to build the new variants around the standard f8 function in such a manner that previous hardware implementations of f8 could be reused. The project was completed in May 2002. As a result of this project the 3GPP standard confidentiality algorithm for f8 was expanded to a family of confidentiality algorithms that could serve different generations and enhancements of mobile systems. The specifications and test data of the algorithms are given in [31]–[33]. The design and evaluation report [34] basically contains the same information as [17].

Cryptographic separation between the different use of the algorithm is achieved by defining a special separation parameter and embedding it in the 26 least significant bits of the initialization value. To support the extension of the 3GPP f8 algorithm for use in GSM, ECSD and GPRS, it is important that the 26 least-significant bits are not hard-coded but left open to take different values.

The specification was to contain three encryption algorithms: A5/3 for GSM and ECSD, GEA3 for GPRS and f8 for UMTS. The common part of all these algorithms was identified and given the name KGCORE (Core Keystream Generator). Then the different interfaces for each particular use were specified for the KGCORE (the notation and other presentation conventions were the same as before).

6.7.2 List of variables

The naming of variables in the specification of the extended f8 differs slightly from the notation used in [19], since it now has to support the GSM and GPRS systems as well. The list of variables is:

BLOCK 1	A string of keystream bits output by the A5/3 algorithm—114 bits for GSM and 348 bits for ECSD.
BLOCK 2	A string of keystream bits output by the A5/3 algorithm—114 bits for GSM and 348 bits for ECSD.
BLOCKS	An integer variable indicating the number of successive applications of KASUMI that need to be performed.
CA	An 8-bit input to the KGCORE function.
CB	A 5-bit input to the KGCORE function.
CC	A 32-bit input to the KGCORE function.
CD	A 1-bit input to the KGCORE function.
CE	A 16-bit input to the KGCORE function.

CK	A 128-bit input to the KGCORE function.
CL	An integer input to the KGCORE function, in the range $1, \ldots, 2^{19}$ inclusive, specifying the number of output bits for KGCORE to produce.
CO	The output bitstream (CL bits) from the KGCORE function.
COUNT	A 22-bit, frame-dependent input to both the GSM and ECSD A5/3 algorithms.
DIRECTION	A 1-bit input to the GEA3 algorithm, indicating the direction of transmission (uplink or downlink).
INPUT	A 32-bit, frame-dependent input to the GEA3 algorithm.
Kc	The cipher key that is an input to each of the three cipher algorithms defined here. Although at the time of writing the standards specify that Kc is 64 bits long, the algorithm specifications here allow it to be of any length between 64 and 128 inclusive, to allow for possible future enhancements to the standards.
KLEN	The length of Kc in bits, between 64 and 128 inclusive (see above).
KM	A 128-bit constant that is used to modify a key. This is used in the KGCORE function.
KS[i]	The ith bit of keystream produced by the keystream generator in the KGCORE function.
KSB_i	The ith block of keystream produced by the keystream generator in the KGCORE function. Each block of keystream comprises 64 bits.
M	An input to the GEA3 algorithm, specifying the number of octets of output to produce.
OUTPUT	The stream of output octets from the GEA3 algorithm.

6.7.3 Core function KGCORE

6.7.3.1 Introduction

In this section, the general-purpose keystream generation function KGCORE is defined. Individual encryption algorithms for GSM, GPRS and ECSD will each be defined in subsequent sections by mapping relevant inputs to the inputs of KGCORE and mapping the output of KGCORE to relevant outputs. After specification of 2G algorithms, an alternative specification of the standard 3GPP f8 algorithm is also given in terms of the KGCORE.

6.7.3.2 Inputs and outputs

Recall (see Section 6.6.2) that the 64-bit input to the f8 algorithm is defined as:

$$\text{COUNT} \parallel \text{BEARER} \parallel \text{DIRECTION} \parallel 0 \ldots 0$$

obtained as the concatenation of the 32-bit COUNT, 5-bit BEARER, 1-bit DIRECTION values and a string of 26 zero bits. In the extension, the f8 field COUNT is redefined as CC, BEARER as CB and DIRECTION as CD. In addition, a new 8-bit field CA is defined. These 8 bits are used to specify the mode of encryption and are taken from the 26 bits that were set equal to 0 in the f8 specification. In addition, a 2-byte field CE is specified as a variable string of bits reserved for possible future uses of KGCORE. The CE field will replace the last two bytes of the all-zero field. All the algorithms specified below assign a constant, all-zero value to CE. After specification of these fields, 2 bits remain unspecified and they are set equal to 0.

The output of KGCORE is the keystream CO of CL bits, denoted by CO[0], ..., CO[CL−1]. The field CL corresponds to what the f8 specification denoted by LENGTH (see also Section 6.7.7.2).

6.7.3.3 Components and architecture

The KGCORE function is based on the KASUMI block cipher, which is specified in [20] (see also Chapter 8). KASUMI is used in a form of OFB mode specified for the f8 algorithm in [19] (see also Section 6.6) and generates the output bitstream in multiples of 64 bits.

6.7.3.4 Initialization

Before generation of keystream bits as output, KGCORE is initialized with the input variables $IV = \text{CC} \parallel \text{CB} \parallel \text{CD} \parallel 0\,0 \parallel \text{CA} \parallel \text{CE}$, KM is set equal to 16 repetitions of the octet $0 \times 55 = 01010101$ and KSB_0 is set to 0. A single operation of KASUMI is then applied to IV, using a modified version of CK, to compute the prewhitening value $W = \text{KASUMI}_{\text{CK} \oplus \text{KM}}(IV)$.

6.7.3.5 Keystream generation

Once the keystream generator has been initialized with the prewhitening value, it is ready to be used to generate keystream bits. The keystream generator produces bits in blocks of 64 at a time, but the number CL of the required output bits may not be a multiple of 64. Between 0 and 63 of the least significant bits are therefore discarded from the last block, depending on the total number of bits specified by CL. The operation of KGCORE is depicted in Figure 6.4, where the static register to store the prewhitening value is denoted by A and the counter to store the synchronization value is denoted by BLKCNT.

Let BLOCKS equal CL/64 rounded up to the nearest integer (e.g., if CL = 128 then BLOCKS = 2 and if CL = 129 then BLOCKS = 3). To generate each KSB the following operation is performed. For each integer n with $1 \leq n \leq$ BLOCKS the nth block of KSB is defined as:

$$\mathrm{KSB}_n = \mathrm{KASUMI}_{\mathrm{CK}}(\mathrm{W} \oplus (n-1) \oplus \mathrm{KSB}_{n-1})$$

The individual bits of the output are extracted from KSB_1 to $\mathrm{KSB}_{\mathrm{BLOCKS}}$ in turn, the most significant bit first, by applying the following operation. For $n = 1, \ldots,$

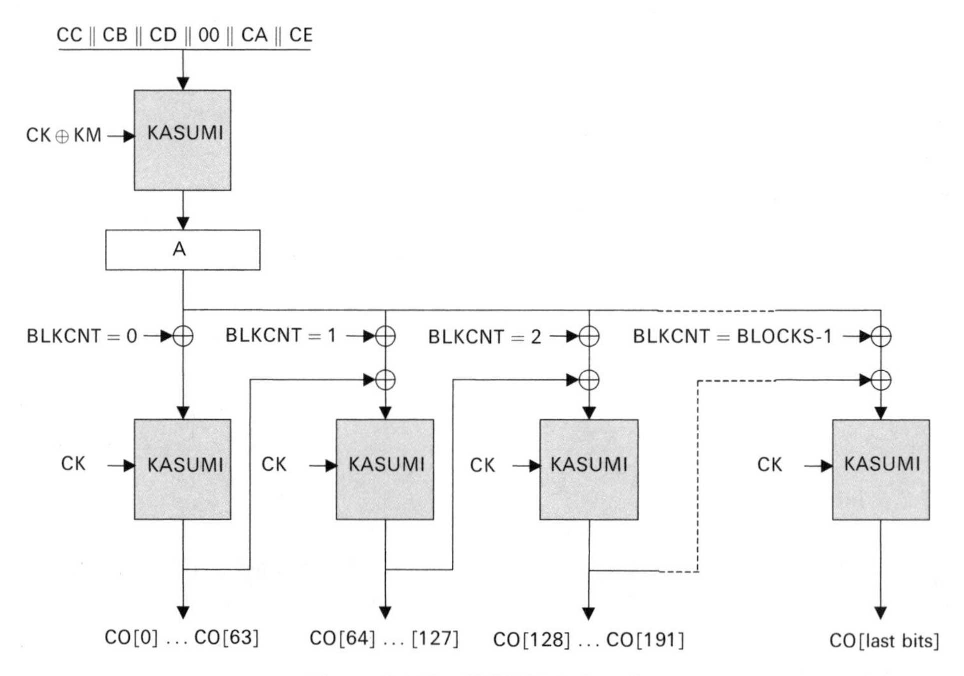

Figure 6.4 The KGCORE function
CK = Confidentiality Key; KM = Key Modifier; BLKCNT = block counter; CO = KGCORE output

BLOCKS and for each integer $i = 0, \ldots, 63$, the keystream sequence CO is defined as:

$$\mathrm{CO}[((n-1)64) + i] = \mathrm{KSB}_n[i]$$

Note that, even if the counter value is specified as a 64-bit integer, it is not necessary to reserve all 64 bits for it in the implementation (see also Section 6.9). A 64-bit variable is used in the description for mathematical reasons so that all operands in the expression $\mathrm{A} \oplus (n-1) \oplus \mathrm{KSB}_{n-1}$ are of the same size.

6.7.4 A5/3 algorithm for GSM encryption

6.7.4.1 Introduction

As defined in [29] the A5 algorithm takes a 64-bit key Kc and a 22-bit counter COUNT and produces two 114-bit keystream blocks (BLOCK1 and BLOCK2) for uplink and downlink traffic. Moreover, as noted in [29], if the actual length of the ciphering key is less than 64 bits, then it is assumed that the actual ciphering key corresponds to the most significant bits of Kc and that the remaining and less significant bits are set to 0. However, it must be clear that for signalling and testing purposes Kc is considered to compose 64 unstructured bits. The GSM A5/3 encryption algorithm, defined in terms of KGCORE, satisfies these external requirements and allows for possible future enhancements to support a longer Kc.

6.7.4.2 Mapping the input and output parameters

The ciphering function is defined by mapping the GSM A5/3 input parameters onto the input parameters of KGCORE and by mapping the output keystream of KGCORE onto the keystream output blocks of GSM A5/3. These mappings are defined as follows:

$$\mathrm{CA}[0]\ldots\mathrm{CA}[7] = 0\,0\,0\,0\,1\,1\,1\,1$$

$$\mathrm{CB}[0]\ldots\mathrm{CB}[4] = 0\,0\,0\,0\,0$$

$$\mathrm{CC}[0]\ldots\mathrm{CC}[9] = 0\,0\,0\,0\,0\,0\,0\,0\,0\,0$$

$$\mathrm{CC}[10]\ldots\mathrm{CC}[31] = \mathrm{COUNT}[0]\ldots\mathrm{COUNT}[21]$$

$$\mathrm{CD}[0] = 0$$

$$\mathrm{CE}[0]\ldots\mathrm{CE}[15] = 0\,0\,0\,0\,0\,0\,0\,0\,0\,0\,0\,0\,0\,0\,0\,0$$

$$\mathrm{CL} = 228$$

Moreover, the CK is extended to a 128-bit-long string as follows. First, it is defined as:

$$CK[0] \ldots CK[KLEN - 1] = Kc[0] \ldots Kc[KLEN - 1]$$

If the Key Length (KLEN) is less than 128, then the remaining bits of CK are filled by repeating the bits of Kc as follows:

$$CK[KLEN] \ldots CK[127] = Kc[0] \ldots Kc[127 - KLEN]$$

So, if KLEN = 64, then CK = Kc || Kc and since KLEN $\geq$ 64 this extension is equivalent to the usual f8 method defined in Section 6.1.3.1.

6.7.4.3 Function definition

The KGCORE function is applied to the inputs given in Section 6.7.4.2 to derive the 228-bit output block CO[0] ... CO[227]. This block is then split into two blocks for uplink and downlink encryption/decryption as follows:

$$BLOCK1[0] \ldots BLOCK1[113] = CO[0] \ldots CO[113]$$

$$BLOCK2[0] \ldots BLOCK2[113] = CO[114] \ldots CO[227]$$

6.7.5 A5/3 algorithm for ECSD encryption

6.7.5.1 Introduction

In ECSD the block size is greater than 114 bits. So, for use in ECSD a modification of the A5 algorithm is employed that produces BLOCK1 and BLOCK2, each containing 348 bits (the input parameters are not modified). It is possible in ECSD for the plaintext data block for either uplink or downlink to be shorter than 348 bits. In this case only the first part of the corresponding output parameter BLOCK is used for bit-wise addition and the remaining bits are discarded.

The ECSD A5/3 algorithm is defined in terms of KGCORE. Similarly to the GSM A5/3 algorithm, the ECSD A5/3 algorithm allows for possible future enhancements to support a longer encryption key Kc.

The ECSD A5/3 algorithm differs from the GSM A5/3 algorithm by using a different algorithm identifier value CA, making it unclear how switching can be implemented between algorithms in Mobile Equipment (ME) that use both ECSD and GSM modulation in parallel. This difficulty can be solved in the following manner: if at least one of the radio channels (uplink or downlink) uses ECSD modulation, then the ECSD A5/3 algorithm can be used for encryption/decryption of the traffic on all channels. If all channels only use GSM modulation, then the GSM A5/3 algorithm can be used.

6.7.5.2 Mapping the input and output parameters

ECSD A5/3 input parameters are mapped onto the input parameters of KGCORE and the output keystream of KGCORE is mapped onto the output blocks of ECSD A5/3 as follows:

$$CA[0]\ldots CA[7] = 1\ 1\ 1\ 1\ 0\ 0\ 0\ 0$$
$$CB[0]\ldots CB[4] = 0\ 0\ 0\ 0\ 0$$
$$CC[0]\ldots CC[9] = 0\ 0\ 0\ 0\ 0\ 0\ 0\ 0\ 0\ 0$$
$$CC[10]\ldots CC[31] = COUNT[0]\ldots COUNT[21]$$
$$CD[0] = 0$$
$$CE[0]\ldots CE[15] = 0\ 0\ 0\ 0\ 0\ 0\ 0\ 0\ 0\ 0\ 0\ 0\ 0\ 0\ 0\ 0$$
$$CL = 696$$

Moreover, the CK is extended to a 128-bit-long string as follows. First, it is defined as:

$$CK[0]\ldots CK[KLEN-1] = Kc[0]\ldots Kc[KLEN-1]$$

If KLEN is less than 128, then the remaining bits of CK are filled by repeating the bits of Kc as follows:

$$CK[KLEN]\ldots CK[127] = Kc[0]\ldots Kc[127-KLEN]$$

So, if KLEN = 64, then CK = Kc || Kc.

6.7.5.3 Function definition

The ECSD A5/3 keystream generator is defined using the KGCORE function, which is applied to the inputs, defined in Section 6.7.5.2, to derive the output CO[0]...CO[695]. To derive the ECSD A5/3 output this block is split into two blocks as follows:

$$BLOCK1[0]\ldots BLOCK1[347] = CO[0]\ldots CO[347]$$
$$BLOCK2[0]\ldots BLOCK2[347] = CO[348]\ldots CO[695]$$

6.7.6 GEA3 algorithm for GPRS encryption

6.7.6.1 Introduction

The external interfaces of the GPRS encryption algorithm GEA are defined in [28], where the interface parameters are defined as follows:

$$\begin{aligned} \text{Kc} &= K[0]\ldots K[63], && \text{where } K[i] \text{ is the } i\text{th bit of Kc} \\ \text{DIRECTION} &= Z[0], && \text{where } Z[0] \text{ is the DIRECTION bit} \\ \text{INPUT} &= X[0]\ldots X[31], && \text{where } X[i] \text{ is the } i\text{th INPUT bit} \\ \text{OUTPUT} &= W[0]\ldots W[M-1], && \text{where } W[i] \text{ is the } i\text{th data output octet} \end{aligned}$$

Uplink and downlink transfers are independent and so encryption for uplink and downlink will also be independent of each other, contrasting with algorithm A5 where keystreams for both directions are generated from the same input. GPRS performance requirements are specified in GSM 02.60 where a distinction is made between a Mobile Station (MS) that admits only one time slot GPRS communication and an MS that admits GPRS communication over the maximum number of eight time slots in both direction.

The performance requirements of the GPRS ciphering algorithm in the first scenario are expected to be similar to the performance of the A5 algorithm. It is also expected that the performance will increase linearly depending on the number of time slots the MS is able to use for GPRS.

The GPRS GEA algorithm produces a keystream string of M octets, but as noted above the number M can vary. Under normal use of the algorithm the data packets to be encrypted are either short packets (25–50 octets) or long packets (500–1,000 octets). The GEA3 specification assumes that M will never exceed $2^{16} = 65{,}536$.

The function of the GEA3 algorithm is defined in terms of KGCORE. Similarly to the GSM A5/3 algorithm, the GEA3 allows for possible future enhancements to support a longer encryption key Kc.

6.7.6.2 Mapping the input and output parameters

The function of the GEA3 algorithm is defined by mapping GEA inputs onto the inputs of KGCORE and mapping the output of KGCORE onto the outputs of GEA as follows:

$$\begin{aligned}
\mathrm{CA}[0]\ldots\mathrm{CA}[7] &= 1\ 1\ 1\ 1\ 1\ 1\ 1\ 1 \\
\mathrm{CB}[0]\ldots\mathrm{CB}[4] &= 0\ 0\ 0\ 0\ 0 \\
\mathrm{CC}[0]\ldots\mathrm{CC}[31] &= \mathrm{INPUT}[0]\ldots\mathrm{INPUT}[31] \quad (= X[0]\ldots X[31]) \\
\mathrm{CD}[0] &= \mathrm{DIRECTION}[0] \quad (= Z[0]) \\
\mathrm{CE}[0]\ldots\mathrm{CE}[15] &= 0\ 0\ 0\ 0\ 0\ 0\ 0\ 0\ 0\ 0\ 0\ 0\ 0\ 0\ 0\ 0 \\
\mathrm{CK}[0]\ldots\mathrm{CK}[63] &= \mathrm{Kc}[0]\ldots\mathrm{Kc}[63] \quad (= K[0]\ldots K[63]) \\
\mathrm{CL} &= 8M
\end{aligned}$$

where the original notation used in [18] is given in parentheses. Moreover, the CK is extended to a 128-bit-long string as follows. First, it is defined as:

$$\mathrm{CK}[0]\ldots\mathrm{CK}[\mathrm{KLEN}-1] = \mathrm{Kc}[0]\ldots\mathrm{Kc}[\mathrm{KLEN}-1]$$

If KLEN is less than 128 then the remaining bits of CK are filled by repeating the bits of Kc as follows:

$$\mathrm{CK}[\mathrm{KLEN}]\ldots\mathrm{CK}[127] = \mathrm{Kc}[0]\ldots\mathrm{Kc}[127-\mathrm{KLEN}]$$

So, if KLEN = 64, then CK = Kc || Kc.

6.7.6.3 Function definition

By applying KGCORE to the inputs defined in Section 6.7.6.2 we derive the output $\mathrm{CO}[0]\ldots\mathrm{CO}[8M-1]$. Then for $0 \le i \le M-1$ the GEA3 output is defined as:

$$\mathrm{OUTPUT}\{i\}(= W[i]) = \mathrm{CO}[8i]\ldots\mathrm{CO}[8i+7]$$

where $\mathrm{CO}[8i]$ is the most significant bit of the octet.

6.7.7 Specification of the 3GPP confidentiality algorithm f8

6.7.7.1 Introduction

The extensions of the 3GPP confidentiality algorithm share the same core function as the 3GPP algorithm. This means that the f8 algorithm can also be viewed in a similar manner to the KGCORE function. Moreover, since all these algorithms are intended to be used within the same UE, it is helpful to give the specification of the f8 algorithm in terms of the KGCORE function to ease simultaneous implementation of multiple algorithms. In this section, a specification of f8 is given in terms of the KGCORE function. However, the definitive specification of f8 remains the one given in [15].

The definitive specification of f8 specifies the entire encryption procedure, which includes not only the generation of the keystream (KS) but also the encryption procedure. Hence the definitive specification of f8 includes an Input Bit Stream (IBS) and an Output Bit Stream (OBS), both of which are LENGTH bits long. OBS is obtained by the bitwise XOR of the IBS and KS. In this section only the keystream generator part of f8 is described for closer comparison with A5/3 and GEA3.

6.7.7.2 Mapping the input and output parameters

The f8 function is defined by mapping the f8 inputs onto the inputs of KGCORE and mapping the output of KGCORE onto the outputs of f8. It is defined as:

$$\begin{aligned}
\mathrm{CA}[0]\ldots\mathrm{CA}[7] &= 0\,0\,0\,0\,0\,0\,0\,0 \\
\mathrm{CB}[0]\ldots\mathrm{CB}[4] &= \mathrm{BEARER}[0]\ldots\mathrm{BEARER}[4] \\
\mathrm{CC}[0]\ldots\mathrm{CC}[31] &= \mathrm{COUNT}[0]\ldots\mathrm{COUNT}[31] \\
\mathrm{CD}[0] &= \mathrm{DIRECTION}[0] \\
\mathrm{CE}[0]\ldots\mathrm{CE}[15] &= 0\,0\,0\,0\,0\,0\,0\,0\,0\,0\,0\,0\,0\,0\,0\,0 \\
\mathrm{CK}[0]\ldots\mathrm{CK}[127] &= \mathrm{CK}[0]\ldots\mathrm{CK}[127] \\
\mathrm{CL} &= \mathrm{LENGTH}
\end{aligned}$$

6.7.7.3 Function definition

The KGCORE function is applied to the inputs given in Section 6.7.7.2 to derive the output $\mathrm{CO}[0]\ldots\mathrm{CO}[\mathrm{LENGTH}-1]$. Then the keystream output (KS) of the f8 can be defined as:

$$\mathrm{KS}[0]\ldots\mathrm{KS}[\mathrm{LENGTH}-1] = \mathrm{CO}[0]\ldots\mathrm{CO}[\mathrm{LENGTH}-1]$$

The operation of the 3GPP confidentiality algorithm f8 is depicted in Figure 6.5.

6.7.8 Summary of the confidentiality functions

To ease the simultaneous implementation of multiple algorithms the input and output parameters of the four algorithms specified above are summarized in Table 6.1.

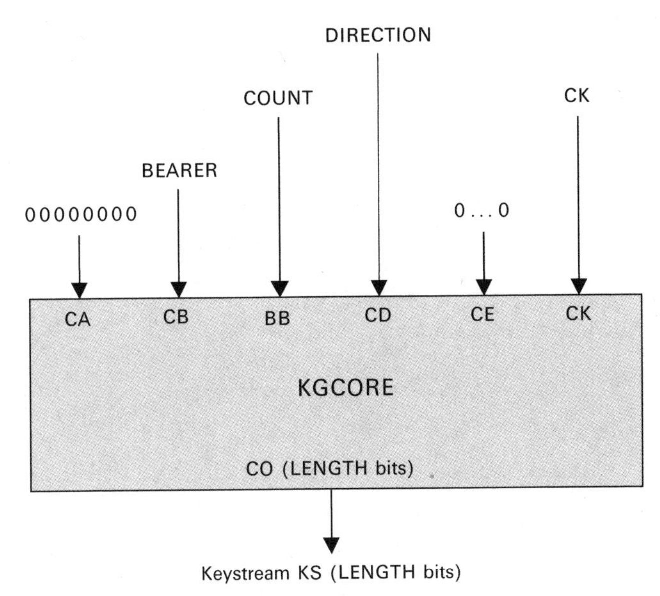

Figure 6.5 3GPP f8 keystream generator function
3GPP = Third Generation Partnership Project; CK = Cipher Key; CA, CB, CC, CD, CE = KGCORE inputs; KGCORE = Core Keystream Generator; CO = KGCORE output

Table 6.1 GSM A5/3 ECSD A5/3, GEA3 and f8 in terms of KGCORE

	GSM A5/3	ECSD A5/3	GEA3	f8
CA	0 0 0 0 1 1 1 1	1 1 1 1 0 0 0 0	1 1 1 1 1 1 1 1	0 0 0 0 0 0 0 0
CB	0 0 0 0 0	0 0 0 0 0	0 0 0 0 0	BEARER
CC	0...0 ‖ COUNT	0...0 ‖ COUNT	INPUT	COUNT
CD	0	0	DIRECTION	DIRECTION
CE	0 0...0	0 0...0	0 0...0	0 0...0
CK	Kc repeated to fill 128 bits	Kc repeated to fill 128 bits	Kc repeated to fill 128 bits	Kc repeated to fill 128 bits
CO	BLOCK1 ‖ BLOCK2	BLOCK1 ‖ BLOCK2	OUTPUT	KS

GSM = Global System of Mobile; ECSD = Enhanced Circuit Switched Data; CA, CB, CC, CD, CE = KGCORE inputs; KGCORE = Core Keystream Generator; CK = Cipher Key; CO = KGCORE output

6.8 Integrity Algorithm

6.8.1 The f9 MAC mode

The 3GPP integrity algorithm f9 computes a 32-bit MAC on an input message under an IK and imposes no limitation on the input message length.

For ease of implementation the algorithm is based on the same block cipher KASUMI as used by the confidentiality algorithm f8. The approach adopted uses

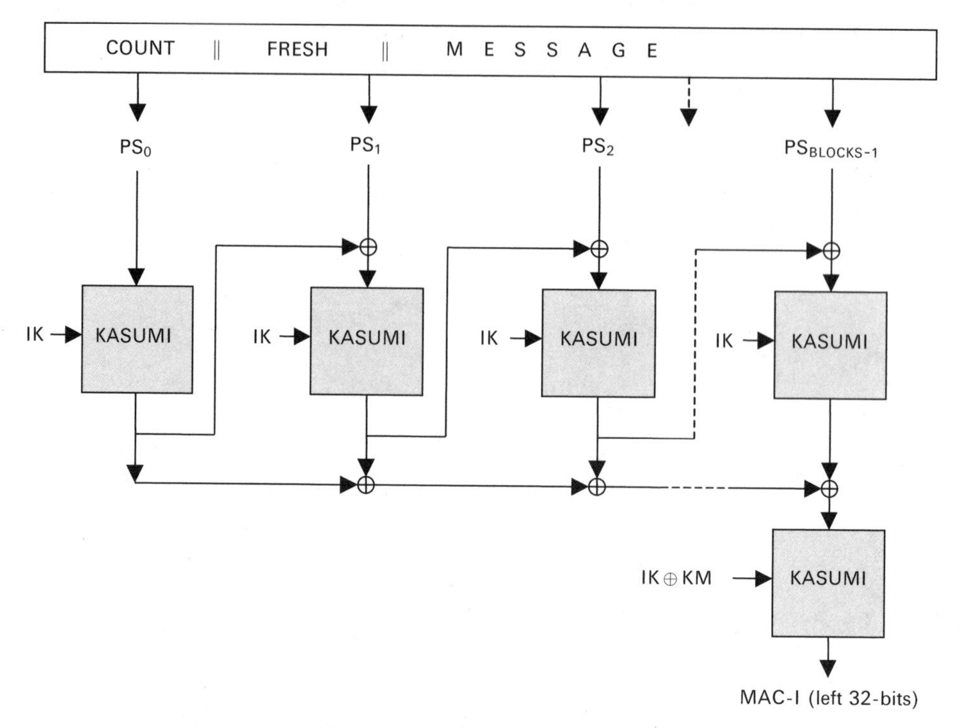

Figure 6.6 The f9 integrity mode
PS = Padded String; IK = Integrity Key; KM = Key Modifier; MAC = Message Authentication Code

KASUMI in a form of CBC-MAC mode, which modifies the standard CBC-MAC mode by adding an operation that combines all intermediate outputs using bitwise XOR and applies one more KASUMI operation on the combined block. The 64-bit output from the last KASUMI operation is then truncated to produce a 32-bit MAC-I.

The operation of the 3GPP standard f9 algorithm is depicted in Figure 6.6.

6.8.2 Description

The 3GPP standard f9 function makes use of two 64-bit registers A and B. The initial value for both registers is set equal to 0: A = 0 and B = 0. The function also makes use of a constant value for a 128-bit KM that is equal to 16 repetitions of the octet 0xAA = 10101010. The inputs to the f9 function are defined in Section 6.2.2: the 32-bit value COUNT, 32-bit value FRESH, a bit string MESSAGE of unlimited length LENGTH and the 1-bit value DIRECTION.

The values of all inputs are concatenated and then a single "1" bit is appended to this string, followed by between 0 and 63 "0" bits, so that the total length of the resulting string is an integer multiple of 64 bits. This string is called the Padded String (PS). Then:

$$PS = COUNT \parallel FRESH \parallel MESSAGE \parallel DIRECTION \parallel 1 \parallel 0 \ldots 0$$

where the number of "0" bits in the padding field is between 0 and 63. The PS is then split into 64-bit blocks PS_i. Let BLOCKS be the number of the resulting 64-bit blocks. Then:

$$PS = PS_0 \parallel PS_1 \parallel PS_2 \parallel \ldots \parallel PS_{BLOCKS-1}$$

This PS is the data input to the MAC algorithm. It would also be possible to interpret the first block PS_0 of PS as the initial value, since $PS_0 = COUNT \parallel FRESH$, and this initial value would be different for each message. For each integer n with $0 \leq n \leq BLOCKS - 1$ the following operations are performed:

$$A = KASUMI_{IK}(A \oplus PS_n)$$
$$B = B \oplus A$$

Finally a further application of KASUMI is performed using a modified form of the IK, as follows:

$$B = KASUMI_{IK \oplus KM}(B)$$

The output from KASUMI has 64 bits: MAC-I comprises the leftmost 32 bits of the result and the rightmost 32 bits are discarded.

6.8.3 Security

6.8.3.1 Internal state and collisions

Arguments supporting the designed mode of operation for the 3GPP integrity function are presented in TR 33.909 [17]. The main reason for not selecting the standard CBC mode was the relatively short block length (64 bits) of KASUMI. In the standard CBC mode, the internal state of the algorithm is equal to the block size of the kernel function, which would mean in the above description that only register A could be used for the computation of the algorithm. With the enhanced construction the internal state is increased to 128 bits comprising the contents of both registers A and B.

If a regular CBC-MAC mode had been chosen for the f9 algorithm, the internal state fed forward from block to block would have only been 64 bits long. In this case a 2^{33}-message birthday attack would likely yield an internal state collision. Having identified a pair of padded strings M and M' for which such a collision occurs, one can always be sure that the padded strings $M \parallel X$ and $M' \parallel X$ have the same MAC for any extension X. In other words, if you can obtain the MAC for $M \parallel X$, then you can forge the MAC for $M' \parallel X$.

This attack would be unrealistic in the 3GPP context, but nevertheless the Task Force preferred the current f9 construction to the standard CBC-MAC mode, because it provides a 128-bit internal state at almost no extra cost. The f9 construction prevents the collision attack with 2^{33} messages, seemingly without introducing any other weaknesses. The straightforward collision attack on this construction requires 2^{65} chosen input data, which is completely out of reach.

6.8.3.2 Knudsen and Mitchell's analysis

It seems that the f9 construction achieved some advantage over the standard CBC-MAC mode. The best-known attack on f9 found by Knudsen and Mitchell [70] requires approximately 2^{48} chosen input messages, which is still considerably more than for the regular CBC-MAC mode.

Knudsen and Mitchell investigate a number of different types of attacks both for key recovery and MAC forgery. The complexity of each attack is determined in terms of the block length n and the final MAC length m. All presented key recovery attacks are infeasible for f9. The best-known attack found by Knudsen and Mitchell is a MAC forgery attack that can be launched if $m < n$. It requires on average $2^{(n+m)/2}$ known data string/MAC pairs and $2^{n-m/2}$ chosen data string/MAC pairs. These numbers are exactly equal if $m = n/2$. For the block size of f9, the numbers of known pairs and chosen pairs required for this attack are equal ($= 2^{48}$) if the MAC length is 32. With shorter MAC lengths the attack would require more chosen input string/MAC pairs, and with longer MAC lengths the number of required known pairs would have been larger. Hence, with respect to this attack, the chosen MAC length seems to offer the weakest security, but it should be kept in mind that independently of the used MAC generation algorithm there is a straightforward MAC forgery attack that requires on average 2^{m-1} online MAC verifications. The MAC length of 32 bits can be seen as a compromise between the straightforward MAC forgery attack and the limited bandwidth resources' over the air interface.

6.8.3.3 Other properties

The Task Force also made the following observations about f9, none of which, however, seems to present any security weakness:

- In the standard CBC-MAC, a change in a single data block will change the MAC, with probability 1. This property does not hold for f9, unless the change is in the last block.
- For every value of the chaining variable A, there exists an input block M such that the output again is A (i.e., $\text{KASUMI}_{\text{IK}}(\text{A} \oplus M) = \text{A}$). Note that both A and M are completely unknown and depend on the value of IK. Then inserting block M an even number of times one after another will cancel the effect in B and, therefore, will not affect the final MAC value.
- As a different consequence of the preceding property, if $\text{A} = 0$ (an event with probability 2^{-64} that cannot be detected easily by an opponent), then inserting M such that $\text{KASUMI}_{\text{IK}}(M) = 0$ (which again is hard to find) an arbitrary number of times will not affect the MAC value.

6.8.3.4 Toward a security proof

The Task Force did not provide any formal security proof for the given construction of the 3GPP MAC algorithm. The function uses a block cipher algorithm as its main security component. The question is how well the good properties of a strong encryption function are transferred to the security properties of a MAC function by the f9 construction. Recall that the security requirement for a MAC algorithm is to prevent any MAC value from being forged even if the attacker is given a large number of valid message and MAC pairs (see Section 4.2.3.1). Forgery is always possible simply by guessing the correct MAC value, but the chances are no better than $1/2^m$. For block ciphers, on the other hand, the strongest security notion is that of indistinguishability (see Section 4.2.1.1), which means that a block cipher with a randomly-chosen key resists all attempts to distinguish it from a randomly chosen permutation.

Recently, Dowon Hong et al. gave a security proof for the 3GPP MAC algorithm in [57].

The proof has not been sufficiently verified yet, but it suggests that under some reasonable assumptions the security of the 3GPP MAC algorithm is at least as good as the security of the standard CBC-MAC with the same size of a block cipher and with the same MAC length.

But could it be strictly stronger as the designers anticipated? Recall that the best known attack by Knudsen and Mitchell (see Section 6.8.3.2) requires 2^{48} chosen input messages, while the CBC-MAC security bound is 2^{32} and the complexity 2^{48} of the best known attack. Establishing the security proof and closing this gap poses an interesting cryptographic research challenge.

6.9 Implementation

6.9.1 Length of data

The standard 3GPP confidentiality algorithm f8 is designed so that, for the given inputs and the secret key, the algorithm produces a keystream sequence of required length. Since the encryption function will be applied in two different layers, the MAC sublayer and the RLC sublayer, consideration may be given to whether simultaneous or interleaving applications of different encryption instances should be supported. If just one implementation of the f8 function is available, then interleaving the execution of two or more encryption instances could in principle be supported. This would require different encryption instances to memorize the internal state of the keystream generator at the time of interruption.

Such a stop-and-go facility can be supported by a special implementation of any keystream generator. However, memory requirements may differ significantly for different types of algorithms. Similarly, the integrity function f9 could have such a stop-and-go facility. The standard f8 and f9 algorithms have not been designed to facilitate implementation in stop-and-go mode. Therefore, if such functionality is required it will take additional effort to implement it and will make the implementation significantly more complex.

Another related question is what should the maximum length of data be to be supported by the implementation of the keystream generator or the integrity function. If the implementation offers a stop-and-go facility, then the operation of the function can easily be made to support any length of data. However, since this is not necessarily the case, the implementation must be designed to handle the requisite length of data.

What are the maximum lengths of data to be handled by the f8 and f9 functions? In early versions of algorithm requirement specifications, message lengths remained unspecified. It was not until Release 4 that an informal note was given in which the length of plaintext to be encrypted, of RLC PDU/MAC SDU size, was stated as about 5,000 bits (the length of signalling data input to the integrity function remained unspecified, however).

Later at the SA WG3 meeting in November 2001, it was decided to formally increase the input message length to 20,000 bits and make the corresponding change to the f8/f9 specification document TS 35.201 [19]. The reasons for the change were as follows:

- the limit of 5,114 bits was too low for integrity protection;
- the limit of 5,114 bits for the confidentiality protection should be changed because the maximum physical layer message can be 20,000 bits.

Increase in the maximum length of the input message may also have implications elsewhere for algorithm implementation. In the informative annex of [19] it is noted that in the description of f8 it is assumed that the three operands—A, BLKCNT and KSB_{n-1}—are assumed to be of equal size (i.e., 64-bit each). In practice, it is not necessary to implement BLKCNT as a 64-bit register. When the keystream generator is required to produce no more that 20,000 bits (i.e., no more than 312 blocks) only the least significant 9 bits of the BLKCNT counter need to be realized.

6.10 Intellectual Property Right (IPR) Issues and Exportability

6.10.1 IPR issues

The position of f8 and f9 algorithms with respect to patents and licensing are defined in the algorithm specification documents. The implementing organization must take note of the special licensing requirements for the KASUMI algorithm, but otherwise the requirements do not differ from what is usual for 3GPP specifications. The requirements quoted in the foreword of [19] are as follows:

> The 3GPP Confidentiality and Integrity Algorithms f8 & f9 have been developed through the collaborative efforts of the European Telecommunications Standards Institute (ETSI), the Association of Radio Industries and Businesses (ARIB), the Telecommunications Technology Association (TTA), the T1 Committee.
>
> The f8 & f9 Algorithms Specifications may be used only for the development and operation of 3G Mobile Communications and services. Every Beneficiary must sign a Restricted Usage Undertaking with the Custodian and demonstrate that he fulfills the approval criteria specified in the Restricted Usage Undertaking.
>
> Furthermore, Mitsubishi Electric Corporation holds essential patents on the algorithms. The Beneficiary must get a separate IPR License Agreement from Mitsubishi Electronic Corporation, Japan.
>
> For details of licensing procedures, contact ETSI, ARIB, TTA or T1.

6.10.2 Exportability

According to the cryptographic algorithm requirements in [10] it is the intention that:

> mobile stations should be free from restrictions on export or use, in order to allow the free circulation of 3G terminals, while network equipment which

embody the algorithms may be expected to come under restrictions. It is however the intention that RNC and AuC which embody such algorithms should be exportable under the conditions of the Wassenaar Arrangement.

The requirements for exportability of the f8 and f9 algorithms were stated by SAGE TF 3GPP in [16] as follows:

Mobile stations will not be controlled according to the Wassenaar arrangement, as long as they are "accompanying their user for the user's personal use". They would also be generally exempted from export control as being: "portable or mobile radiotelephones for civil use that are not capable of end-to-end encryption". The intended network wide encryption specified in the 3G architecture could possibly be debated but as it is only allowing network controlled key management it seems it would not qualify as true end-to-end encryption. The mobile stations are thus assumed to fulfil requirements according to [10], as long as the exporting countries abide by the Wassenaar rules.

Network equipment embodying algorithms should be expected to need export control licences according to the present Wassenaar arrangement (December 1998), very much like e.g. base stations for GSM have been and are export controlled today. The SAGE Task Force sees no reason to believe that any special problems should arise in this area which could endanger the fulfilment of requirements for a wide international spread of 3G systems. The SAGE Task Force has, however, no possibilities to guarantee such a situation as the actual export licenses are handled individually by each country (or possibly internationally co-ordinated as by the European Union).

To some extent this topic was also discussed informally with a number of export control authorities and no adverse reactions to these interpretations were announced. It has also been noted by the SAGE Task Force that several countries have introduced more liberal rules than the Wassenaar arrangement indicates, especially in the area of so-called mass market products, which the SAGE Task Force believes could even more alleviate the free movement of mobile stations.

7

Kernel Algorithm KASUMI

7.1 Introduction

The modes of operation f8 and f9 developed by SAGE TF 3GPP were not intended to become general-purpose confidentiality and integrity modes of operation for a general block cipher. They were designed for specific use in the context of UMTS. They were also designed with a specific block cipher algorithm in mind, which was chosen as a starting point for what was going to be the kernel algorithm, a modified version of MISTY1 (see Section 6.4.2). In parallel with the development of the f8 and f9 modes of operation, adjustments were also made to the block cipher algorithm. The final version of the block cipher algorithm is known as KASUMI—*kasumi* is Japanese for "hazy, dim, blurred" (see Figure 7.1). Similarly to the modes around it, the KASUMI kernel function is not meant for use outside its intended application in UMTS. In addition to the cryptographic limitations, the use of KASUMI is also subject to license, which is announced to be granted royalty-free for use in standard UMTS confidentiality and integrity algorithms.

The development of the name of the kernel function, KASUMI, was also an iterative process. The modified MISTY1 was drafted in late August 1999 and was

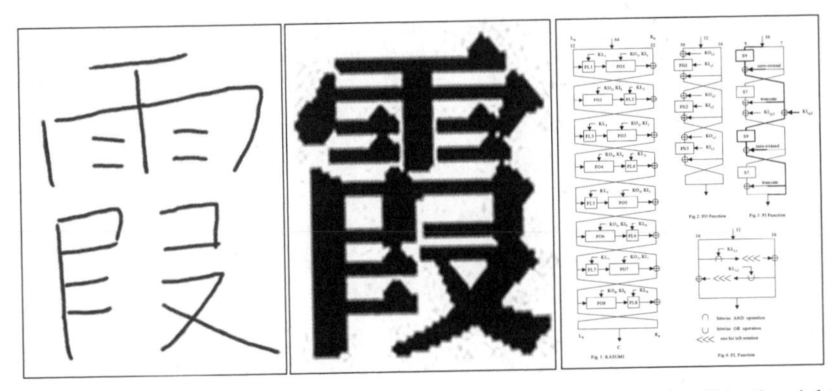

Figure 7.1 Three representations of KASUMI (see Figure 7.3 for the detail in the rightmost part of this figure)

presented to the evaluation team for checking out before the next Task Force meeting in late September. At that time the draft design was referred to using such names as MM1, Modified MISTY1 or M^2 (these names could also have been derived from the initials of the principal designer of the algorithm: Mitsuru Matsui). At the September meeting a new draft was already available. This algorithm was a further modification of MM1 and given the name "FOGGY", which SAGE changed further to "KASUMI".

7.2 MISTY Block Cipher Algorithms

7.2.1 Design principles of MISTY1

7.2.1.1 Resistance to linear and differential cryptanalysis

The foundations of MISTY-type structures were laid by Mitsuru Matsui in [82]. These constructions were motivated by recent developments in the cryptanalysis of block ciphers and more notably by theoretical results about how to achieve resistance to new cryptanalytic methods, such as differential and linear cryptanalysis. Matsui himself was a pioneer of the linear cryptanalysis method [81], which succeeded the invention of the differential cryptanalysis method by Eli Biham and Adi Shamir a few years earlier [41]. It is unclear how well these methods had been known in earlier, secret cryptanalytic work. According to Don Coppersmith [46], the designers of the DES algorithm were aware of differential cryptanalysis, but no such statement had ever been made of linear cryptanalysis. Therefore, it is all the more interesting to note that, of these two methods, linear cryptanalysis is more powerful when applied to the DES.

Ciphers are designed to provide robust unpredictability: given a plaintext it should be very hard to predict what the ciphertext will be after application of the encryption transformation with a fixed but unknown key. The basic idea behind differential and linear cryptanalysis is to consider some other data derived from the plaintext, not just the plaintext itself, and see how accurately the result after encryption could be predicted. The basic differential method considers differences (or XOR sums) between pairs of plaintext blocks. Higher-order differential cryptanalysis is concerned with XOR sums of all plaintext blocks in a small, linear subspace. Linear cryptanalysis analyses the predictability of linear combinations of plaintext bits.

In differential cryptanalysis, the prediction relationship of the difference between a pair of inputs and the difference between the corresponding outputs is called a differential. The strength of a differential is measured by a probability measure taken over all inputs. Biham and Shamir determined all differentials over one round of the DES cipher. Then they selected the strongest and undertook an exhaustive search of all possible ways of combining strong, one-round differentials to a chain that would extend itself over more rounds, and finally from the first round up to the last round. Such a chain is called "characteristic". The strength of the characteristic is measured as the product of the probabilities of one-round differentials it consists of. Moreover, the DES cipher has the property that its characteristic probabilities are independent of the used key. This holds for all block ciphers where the key is XORed to the data between rounds. One such example is MISTY1 without FL functions (see Section 7.2.1.2). The basic differential attack is a chosen plaintext attack, which is used to determine the round keys applied at the first and the last round. Further developments of the differential cryptanalysis method include such concepts as higher-order differentials and impossible differentials, which can be used to distinguish a block cipher from a random permutation, or even to derive bits of the used key.

An important observation made by Xuejia Lai et al. [76] was that the probability of a characteristic derived round by round was, in general, just a lower bound of the probability of the differential over multiple rounds of the iterated block cipher. It meant that the actual strength of a predictive relation between a difference at input and a difference in output could be higher than what was possible to estimate using characteristics. Indeed, all possible characteristics that predict a fixed output difference given a fixed input difference contribute to the differential that predicts this output difference from a given input difference. Unfortunately, it is impossible to consider them all and, therefore, it was necessary to find other approaches to estimate the probability of differentials.

In [95] Nyberg and Knudsen investigated differentials over a DES-like cipher with a general, unspecified round function. Their main observation was that it is sufficient to consider only four rounds of such a cipher. The average probability of a differential over more than four rounds always had an upper bound provided by the average probability of a differential over four rounds, when the average is taken over all round keys. They showed that it was possible to reach the optimal upper bound in four rounds, in which case adding more rounds would not bring any significant improvement to average probability. They also presented a mathematical construction of a round function for which the differential probabilities after four rounds were very close to the theoretical minimum value. It was also shown that the number of rounds to achieve the optimal upper bound can be reduced to three in the case of an invertible round function and that the upper bound of the differential probability was smaller.

The basic idea behind the linear cryptanalysis method invented by Matsui [81] is very similar to that of differential cryptanalysis. Instead of differences between plaintext blocks, linear cryptanalysis works with linear combinations of data bits. Such a linear combination has a value 0 or 1. Given a linear combination of plaintext data one can ask how well its value predicts the value of some linear combination of the ciphertext bits after encryption. In other words, the question is about correlation between a certain linear combination of plaintext bits and a certain linear combination of ciphertext bits. One of the main differences between linear and differential cryptanalysis is that the former only requires the plaintext to be known, while for the latter a specifically chosen plaintext is needed. A theory about cryptanalytic resistance was also developed for the linear cryptanalysis method [48], [94].

7.2.1.2 The nested structure

As reported in [82] the design of MISTY was motivated by the desire to achieve resistance to linear and differential cryptanalysis. The block cipher should also be practical, and Matsui was not pleased by the example given in [95], because "its computational complexity is not small because it requires a calculation over GF(2^{33})" [82]. Matsui was not aware of a more serious drawback of the example cipher [95]: it is based on computing third powers in GF(2^{33}) and because of this its encryption function can be expressed as a polynomial of low degree. The lower the degree of a polynomial the less the number of its coefficients. Hence, a small number of plaintext/ciphertext pairs is sufficient to determine a global reduction of this cipher [64].

Matsui's idea for his construction of a block cipher was original and novel: a nested structure of iterated block ciphers (shown in Figure 7.2). Nyberg and Knudsen had shown that using a differentially resistant function of block length n, it is possible to double the block length using the DES-like structure and obtain a bigger differentially resistant cipher. Matsui adopted this principle and used it three times recursively starting from small block lengths and increasing it three times to achieve a block length of 64 bits. An easy calculation shows that if the block length is doubled at each recursive step, the block length should start with 8 bits. But in the smallest dimension Matsui wanted to use the best possible function, which would have the least possible predictability for differences and linear combinations. The

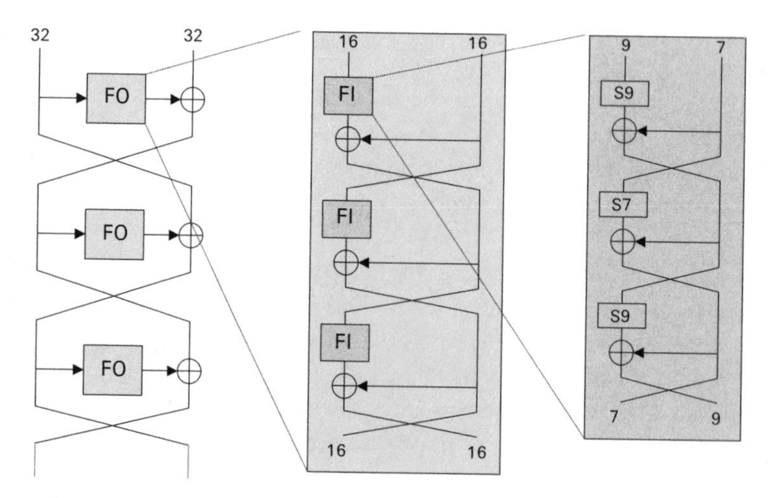

Figure 7.2 The nested structure of MISTY1 (without FL functions)

cubic function has optimal properties, but unfortunately it is not bijective (invertible) in GF(2^8), which is the space where 8-bit blocks reside. The cubic function in GF(2^8) maps three points to one and, therefore, is unsuitable for cipher construction. It is also known that in odd dimensions bijective functions can achieve better nonlinearity (linear unpredictability) than in even dimensions; so, Matsui decided to use dimensions 7 and 9. Then, he selected highly nonlinear bijective functions S7 and S9 in GF(2^7) and GF(2^9), respectively, and built around them a network that produced a function FI from 16 bits to 16 bits. Further, Matsui proved an analogous result to the theorem of [95] that showed that the new type of network can be used to construct larger differentially and linearly resistant ciphers from such smaller functions.

The second step in constructing MISTY was to take three 16 to 16-bit functions FI and combine them using a similar network in such a way that the resulting function is a 32 to 32-bit bijective function, with sufficiently small upper bounds to linear and differential predictability. This function FO is then taken as a round function for a DES-like cipher.

7.2.1.3 Efficient implementation

In addition to providing high resistance to linear and differential cryptanalysis, the functions S7 and S9 where chosen to admit as efficient an implementation in hardware as possible. In software, such small functions are usually implemented as look-up tables, but in hardware they are computed using Boolean gates, such as AND, XOR, NAND and OR gates. The number of Boolean gates to express a function is often referred to as the gate complexity of the function. When searching for suitable functions, Matsui exploited the known fact that variations of functions obtained by linear transformations of the input space, or the output space, have equally as good properties against differential and linear cryptanalysis as the original functions. But the gate complexity of such functions may differ significantly. Therefore, when constructing S7 and S9 Matsui selected a simple power function, known to provide optimal linear and differential unpredictability, as his starting point. Then he applied linear transformations to it, derived the gate representations of the transformed functions and determined their complexity in hardware. Hardware complexity was evaluated based on the *algebraic-normal-form* representation of the output bits of the function. The hardware length of a component function was defined as the total number of XOR gates in the algebraic-normal-form representation plus one. The total hardware length of the function was defined as the maximum hardware length of its components. Matsui's selection for S7 was a function with minimal total hardware length 13 and was obtained by linear variation of the power function x^{13}.

One more property taken into account was the algebraic degree of the function. The algebraic degree of a Boolean function is the degree of the polynomial formed by its algebraic-normal-form representation. Algebraic degree is also used as a measure of nonlinearity and has importance as far as resistance to higher-order differential cryptanalysis is concerned. The higher the algebraic degree the better the resistance. When selecting S7, Matsui could afford a function of algebraic degree 3. The total hardware length for degree 4 functions would have been at least 21 and was considered too expensive. When selecting S9, he had to content himself with algebraic degree 2, which gave him a total hardware length of 12.

The cryptographic concepts of differential and linear unpredictability are closely related to other concepts such as weight distribution and covering radius of error correcting codes. A number of (in this sense) robust functions are known to exist [40], [93]. Special attention was focused on power functions of the form x^{α} in a finite field GF(2^n). The potential applications of such functions in cryptography also inspired mathematicians Hans Dobbertin [50] and Tor Helleseth [55] to research this area.

Finally, to upset the clean and streamlined MISTY structure, Matsui equipped the ciphers with key-dependent simple transformations, known as FL transformations. For each fixed key, FL transformations are linear, but dependence on the key is nonlinear. Because of linearity for a fixed key the FL functions do not disturb the proven resistance to linear and differential cryptanalysis. In MISTY1 the FL functions are applied at every odd round of the outer cipher network (on the left in Figure 7.2) to both 32-bit data block halves.

7.2.2 Security of MISTY

One of the main reasons SAGE selected MISTY1 was that it had already been publicly scrutinized to some extent. Three years had passed since Matsui published the nested construction, two years since MISTY specifications had been given and three cryptanalytic papers had been published [128], [132] and [133]).

The first paper [128] deals with the pseudorandomness of MISTY2. The main difference between MISTY1 and MISTY2 lies in their outer structure: MISTY1 uses the new network structure developed by Matsui for the FI and FO functions, while the outmost structure is the well-known Feistel network; and MISTY2 uses the new network structure at all levels. The advantage of the new structure is that it offers the possibility of parallel computation of round functions at two consecutive rounds. The penalty of this advantage is less efficiency in providing general pseudorandomness properties, as shown in [128]. Mainly due to this observation, MISTY1 became the more successful of the two designs.

The other two paper's ([132] and [133]) investigated the resistance of MISTY ciphers to higher-order differential cryptanalysis. These investigations were theoretical in the sense that the ciphers were simplified by removing the FL functions and reducing the number of rounds to five. Recall that the FL functions do not affect proven resistance to linear and differential cryptanalysis. According to Matsui's proofs, all differentials and linear approximations over three rounds of simplified MISTY1 had very small probabilities. Hence it would be impossible to launch any successful cryptanalytic attack over five rounds of simplified MISTY1 using such differentials or linear approximations. Nevertheless, other types of attacks might exist. In [132] and [133] it was shown that there was a higher-order differential relation with probability 1 over three rounds of MISTY1 without FL functions. Let us give a brief description of this analysis.

While the usual differential cryptanalysis deals with propagation of XOR sums of a pair of plaintext blocks, higher order differential cryptanalysis considers propagation of XOR sums over a linear subspace of the plaintext space. The main discovery ([132] and [133]) was a higher order differential of order 7 (this is the dimension of the subspace) over three rounds of simplified MISTY1. This can be exploited in a chosen plaintext attack over five rounds of the cipher to find a part of the key. The subspace V of the plaintext space of concern is formed by plaintext blocks of the form $(0_{57} \parallel x)$, where 0_{57} denotes a block of 57 zero bits and x is a block of 7 bits. Now, consider the simplified MISTY1 of three rounds. This is almost exactly the kind of network depicted in Figure 7.2, the only difference being that round keys are added using bitwise XOR to the data blocks before the data are taken to the S7 and S9 transformations in the rightmost network. Now, fix any key K and consider the encryption function defined by Figure 7.2. The function that maps the plaintext to the block of the seven leftmost bits of the output is denoted by f_k. Then, it turns out that, for all 64-bit blocks w:

$$\bigoplus_{x \in V} f_K(x \oplus w) = c$$

where c is a constant that is independent of the key K. Such a relation reveals information about the component functions of f_K. Considered as functions of x, these components are polynomials of algebraic degree at most 7. The derived higher-order differential means that the coefficient of the highest-degree term is a constant, independent of the key K.

Now assume that the cipher under consideration is a simplified MISTY1 with five rounds. The derived higher-order differentials give important information about the intermediate values after the third round of the cipher, where an outsider without knowledge of the secret key should otherwise have no access. It was shown in [133] that 11 such relations, with different w values, are sufficient to determine a large part of the last round key. This means that 11×2^7 chosen plaintexts are needed and 2^{17} applications of the FO function for the attack to be successful.

Higher-order differentials only occur for simplified MISTY1 when there is a reduced number of rounds. They are not known to pose any threat whatsoever to the complete MISTY1 with FL functions. Nevertheless, since this property was partly due to the regular structure of MISTY1, it was one of the reasons a fourth round was added to the FI function of KASUMI (as described in the next section).

7.3 Changes between MISTY1 and KASUMI

This section summarizes the changes that were made to MISTY1 during the design of KASUMI. First, the key schedule of MISTY1 was rather complicated. A key-scheduling algorithm of a block cipher is used to derive a number of subkeys for the algorithm to be used at different rounds of the cipher. The subkeys must be recomputed if the encryption key is changed. Since the ciphering keys in the UMTS system are renewed at each authentication process, the desire was to make the computation process as fast as possible.

Some changes were due to previous cryptanalytical results of MISTY1, which had revealed some unwanted regularities in the data encryption function. A number of changes were due to the findings of the Task Force during the design and evaluation process. A detailed account of the changes is given in TR 33.909 [17].

7.3.1 Changes to the data encryption part

Changes to the encryption function of MISTY1 were the following:

1. The location of the FL functions was changed, making hardware simpler but a bit slower. However, this drawback was compensated by other changes. Note that this structure did not block the parallel computation of two FI functions.
2. The subkey $KO_{i,4}$ was removed in the FO function, making hardware simpler and faster because the FO function then had a simple, repetitive structure.
3. Rotate shift functions were added in the FL function. It was assumed that this makes cryptanalysis harder and has no negative impact on hardware size and speed.
4. The substitution table for the function S7 was changed. This was not a significant change; in fact, it was equivalent to just rearranging the bit order before and after the original S7. No better alternative from the viewpoint of hardware implementation had been found.
5. The substitution table for the function S9 was changed, making hardware smaller (and possibly faster). The total number of "terms" of the new S9 in its algebraic normal form was smaller than that of the original S9. A search was made over all polynomials and normal bases, all powers whose Hamming weight is 2 and all linear transformations of the output coordinates for shorter component functions, where the length of the component is defined as the number of terms (except a constant value) in its algebraic normal form (see Section 7.2.1.3). For the new S9, the average length of the component logic is 11.2, while for the S9 of MISTY1 it is 11.7.

6. A second application of S7 was added to the FI function, making the security level significantly higher but hardware bigger. This increase was compensated by the reduction of the key scheduling part. Note that the penalty on hardware speed was not particularly significant because S9 and S7 can be performed in parallel.

7.3.2 Changes to the key-scheduling part

The derivation of subkeys was completely changed. Indeed, all nonlinear functions were replaced by rotations and addition of constants. Even if related key attacks were not considered possible in the UMTS applications of the algorithm, some care was still taken to prevent from such attacks. The changes were as follows:

1. All FI functions were removed in the key scheduling part, making hardware smaller and/or reducing key set-up time. It was expected that related key attacks would not work for this structure.
2. Constant values and rotate shift operations were added, avoiding use of the same values in different rounds.

7.4 Description of KASUMI

7.4.1 General structure

KASUMI is a Feistel cipher with eight rounds. It operates on a 64-bit data block and uses a 128-bit key. The round function (or f-function) used in the ith round of the Feistel cipher is denoted by f_i. The f-function has a 32-bit input and a 32-bit output. Each f-function of KASUMI is composed of two functions: an FL-function and an FO-function. An FO-function is defined as a network that makes use of three applications of an FI-function. An FI-function has a 16-bit input and a 16-bit output. Each FI-function comprises a network that makes use of two applications of a function S9 and two applications of a function S7. The functions S7 and S9 are also called "S-boxes of KASUMI". In this manner KASUMI has a similar three-layer nested structure to MISTY1. For an outline of the MISTY1 encryption function see Figure 7.2. The detailed structure of the KASUMI encryption function is depicted in Figure 7.3.

In this manner KASUMI decomposes into a number of subfunctions (FL, FO and FI) that are used in conjunction with associated subkeys (KL, KO and KI).

The outmost Feistel network comprises eight rounds, which are called in the specification outer rounds and numbered using index i, $i = 1, 2, \ldots, 8$. The

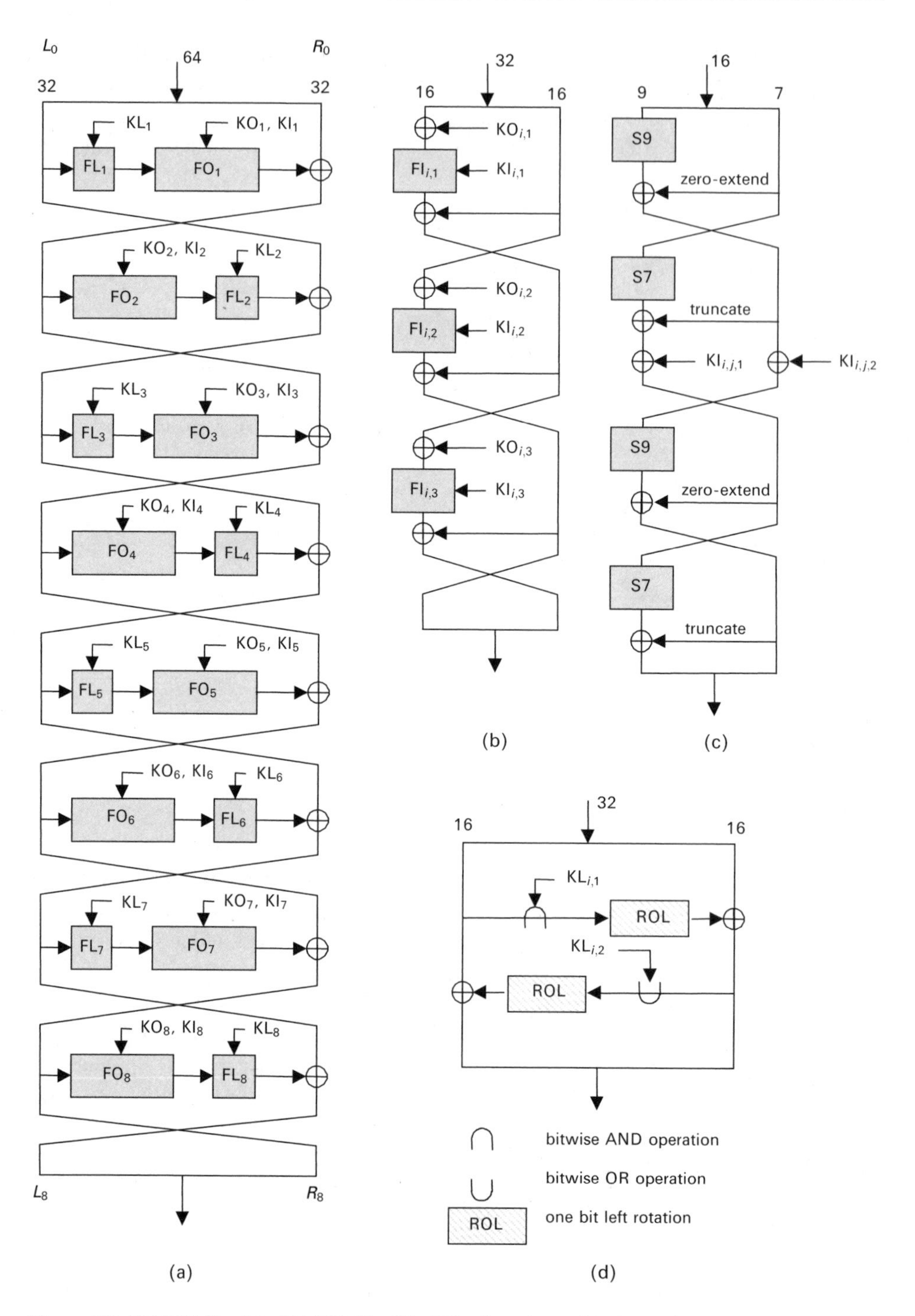

Figure 7.3 KASUMI: (a) KASUMI; (b) FO_i function; (c) $FI_{i,j}$ function and (d) FL_i function

FL-functions and FO-functions used at each round of the Feistel network are numbered accordingly (i.e., FL_i and FO_i are the functions used at the ith round of the outer network). Function FL_i is used in conjunction with subkey KL_i, and function FO_i is used in conjunction with two subkeys: KO_i and KI_i.

The networks formed by the eight FO-functions are called the inner networks and each one has three rounds indexed by j, $j = 1, 2, 3$. Each round of an inner network makes use of a KO-key and an FI-function, the latter is used in conjunction with a KI-key. Consider the ith inner network FO_i. The KO-key, the FI-function and the KI-key used at the jth round of FO_i are denoted as $KO_{i,j}$, $FI_{i,j}$ and $KI_{i,j}$, respectively (e.g., the FI-function used in the third round of the FO-function in the fifth round of KASUMI is denoted as $FI_{5,3}$). In addition, the KI-key $KI_{i,j}$ splits into two halves $KI_{i,j,1}$ and $KI_{i,j,2}$.

7.4.2 KASUMI encryption function

In f8 and f9 mode operation, the kernel function is only computed in one direction. So, even if the kernel function is a block cipher, the decryption transformation is never used. Hence for the purposes of 3GPP only the encryption function of KASUMI needs to be defined. Clearly, it would be possible to derive the definition of the decryption function of KASUMI from the definition of its encryption function, but it is not needed in the 3GPP context.

The fact that KASUMI decryption transformation is never used also explains the difference between the DES encryption function and the KASUMI encryption function. Both algorithms are based on a Feistel network, but the last rounds of their encryption functions are different. For KASUMI the data halves are swapped, while for DES they are not. If the data halves are not swapped at the last round of a Feistel network, then the encryption and decryption transforms are similar, which has a certain advantage for implementation.

7.4.2.1 Outer Feistel network

KASUMI operates on a 64-bit input (INPUT) using a 128-bit key (K) to produce a 64-bit output (OUTPUT), as follows. INPUT is divided into two 32-bit strings L_0 and R_0, where:

$$\text{INPUT} = L_0 \parallel R_0$$

Then for each integer i with $1 \leq i \leq 8$ the operation on the ith round of KASUMI is defined as:

$$R_i = L_{i-1}, \qquad L_i = R_{i-1} \oplus f_i(L_{i-1}, \text{RK}_i)$$

where $L_{i-1} \parallel R_{i-1}$ is the input data block, $L_i \parallel R_i$ is the output data block and RK_i is the ith round key, defined as a triplet of subkeys (KL_i, KO_i, KI_i). The subkeys are derived from the key K using the key-scheduling algorithm given in Section 7.4.3. The output data block (OUTPUT) is defined as:

$$\text{OUTPUT} = L_8 \,\|\, R_8$$

which is the data block offered at the end of the eighth round. In the specification of f8 and f9 this transformation is also denoted as:

$$\text{OUTPUT} = \text{KASUMI}_K[\text{INPUT}]$$

7.4.2.2 f-functions

Each f-function f_i takes a 32-bit input I and returns a 32-bit output O under the control of a round key RK_i, where the round key comprises the triplet (KL_i, KO_i, KI_i). The f-function f_i itself is constructed from two subfunctions: an FL-function FL_i and an FO-function FO_i with associated subkeys KL_i (used with FL_i) and subkeys KO_i and KI_i (used with FO_i).

The f-function f_i has two different forms depending on whether it is an even round or an odd round. For odd rounds $i = 1$, 3, 5 and 7, the f-function is defined as:

$$f_i(I, RK_i) = FO_i(FL_i(I, KL_i), KO_i, KL_i)$$

and for even rounds $i = 2$, 4, 6 and 8, the f-function is defined as:

$$f_i(I, RK_i) = FL_i(FO_i(I, KO_i, KI_i), KL_i)$$

(i.e., for odd rounds first the FL-function and then the FO-function is applied to the round data, while for even rounds the order of the functions is changed).

7.4.2.3 FL-functions

The input to function FL_i comprises a 32-bit data input I and a 32-bit subkey KL_i. The subkey is split into two 16-bit subkeys, $KL_{i,1}$ and $KL_{i,2}$, where:

$$KL_i = KL_{i,1} \,\|\, KL_{i,2}$$

The input data I is split into two 16-bit halves, L and R, where $I = L \,\|\, R$. The FL-functions make use of the following simple operations:

$\text{ROL}(D)$	the left circular rotation of a data block D by one bit
$D_1 \cup D_2$	the bitwise OR operation of two data blocks D_1 and D_2
$D_1 \cap D_2$	the bitwise AND operation of two data blocks D_1 and D_2

Then the 32-bit output value of the FL-function is defined as $L' \| R'$, where:

$$L' = L \oplus \mathrm{ROL}(R' \cup \mathrm{KL}_{i,2})$$
$$R' = R \oplus \mathrm{ROL}(L \cap \mathrm{KL}_{i,1})$$

7.4.2.4 FO-functions

The input to function FO_i comprises a 32-bit data input I and two sets of subkeys: a 48-bit KO_i and 48-bit KI_i. The 32-bit data input is split into two halves, L_0 and R_0, where $I = L_0 \| R_0$, while the 48-bit subkeys are subdivided into three 16-bit subkeys, where:

$$\mathrm{KO}_i = \mathrm{KO}_{i,1} \| \mathrm{KO}_{i,2} \| \mathrm{KO}_{i,3} \qquad \text{and} \qquad \mathrm{KI}_i = \mathrm{KI}_{i,1} \| \mathrm{KI}_{i,2} \| \mathrm{KI}_{i,3}$$

For each integer j with $1 \leq j \leq 3$ the operation of the jth round of the function FO_i is defined as:

$$R_j = \mathrm{FI}_{i,j}(L_{j-1} \oplus \mathrm{KO}_{i,j}, \mathrm{KI}_{i,j}) \oplus R_{j-1}$$
$$L_j = R_{j-1}$$

Output from the FO_i function is defined as the 32-bit data block $L_3 \| R_3$.

7.4.2.5 FI-functions

The FI-functions are depicted in Figure 7.3. The thick and thin lines in this diagram are used to emphasize the difference between the 9-bit and 7-bit data paths, respectively.

An FI-function $\mathrm{FI}_{i,j}$ takes a 16-bit data input I and a 16-bit subkey $\mathrm{KI}_{i,j}$. The input I is split into two unequal components, a 9-bit left half L_0 and a 7-bit right half R_0, where $I = L_0 \| R_0$. Similarly, the key $\mathrm{KI}_{i,j}$ is split into a 7-bit component $\mathrm{KI}_{i,j,1}$ and a 9-bit component $\mathrm{KI}_{i,j,2}$, where $\mathrm{KI}_{i,j} = \mathrm{KI}_{i,j,1} \| \mathrm{KI}_{i,j,2}$.

Each FI-function $\mathrm{FI}_{i,j}$ uses two S-boxes: S7 which maps a 7-bit input to a 7-bit output and S9 which maps a 9-bit input to a 9-bit output. The definition of S-boxes in Section 3.4.4.2.6. FI-functions also use two additional functions, which are designated by ZE and TR. These simple functions are defined as follows:

$\mathrm{ZE}(D)$ takes a 7-bit data string D and converts it to a 9-bit data string by appending two zero bits to the most significant end of D

$\mathrm{TR}(D)$ takes a 9-bit data string D and converts it to a 7-bit value by discarding the two most significant bits of D

The function $\mathrm{FI}_{i,j}$ is defined by the following series of operations:

$$
\begin{array}{ll}
L_1 = R_0 & R_1 = \mathrm{S9}[L_0] \oplus \mathrm{ZE}(R_0) \\
L_2 = R_1 \oplus \mathrm{KI}_{i,j,2} & R_2 = \mathrm{S7}[L_1] \oplus \mathrm{TR}(R_1) \oplus \mathrm{KI}_{i,j,1} \\
L_3 = R_2 & R_3 = \mathrm{S9}[L_2] \oplus \mathrm{ZE}(R_2) \\
L_4 = \mathrm{S7}[L_3] \oplus \mathrm{TR}(R_3) & R_4 = R_3
\end{array}
$$

The output of the $\mathrm{FI}_{i,j}$ function is the 16-bit data block $L_4 \parallel R_4$.

7.4.2.6 S-boxes

The two S-boxes have been designed so that they may be easily implemented in combinational logic or by a look-up table. Both forms are given for each S-box. The input x comprises either seven or nine bits with a corresponding number of bits in the output y. Therefore:

$$x = x_8 \parallel x_7 \parallel x_6 \parallel x_5 \parallel x_4 \parallel x_3 \parallel x_2 \parallel x_1 \parallel x_0$$

and

$$y = y_8 \parallel y_7 \parallel y_6 \parallel y_5 \parallel y_4 \parallel y_3 \parallel y_2 \parallel y_1 \parallel y_0$$

where the x_8, y_8 and x_7, y_7 bits only apply to S9 and the x_0 and y_0 bits are the least significant bits.

For brevity, the specification uses the following conventions in the gate logic equations: for any two bits u and v, the logical AND operation of u and v is denoted by uv. The XOR operation of u and v is designated by $u \oplus v$ and the gate logic of S-box S7 is as follows:

$$
\begin{aligned}
y_0 &= x_1x_3 \oplus x_4 \oplus x_0x_1x_4 \oplus x_5 \oplus x_2x_5 \oplus x_3x_4x_5 \oplus x_6 \oplus x_0x_6 \oplus x_1x_6 \oplus x_3x_6 \\
&\quad \oplus x_2x_4x_6 \oplus x_1x_5x_6 \oplus x_4x_5x_6 \\
y_1 &= x_0x_1 \oplus x_0x_4 \oplus x_2x_4 \oplus x_5 \oplus x_1x_2x_5 \oplus x_0x_3x_5 \oplus x_6 \oplus x_0x_2x_6 \oplus x_3x_6 \oplus x_4x_5x_6 \oplus 1 \\
y_2 &= x_0 \oplus x_0x_3 \oplus x_2x_3 \oplus x_1x_2x_4 \oplus x_0x_3x_4 \oplus x_1x_5 \oplus x_0x_2x_5 \oplus x_0x_6 \oplus x_0x_1x_6 \oplus x_2x_6 \\
&\quad \oplus x_4x_6 \oplus 1 \\
y_3 &= x_1 \oplus x_0x_1x_2 \oplus x_1x_4 \oplus x_3x_4 \oplus x_0x_5 \oplus x_0x_1x_5 \oplus x_2x_3x_5 \oplus x_1x_4x_5 \oplus x_2x_6 \\
&\quad \oplus x_1x_3x_4x_6 \\
y_4 &= x_0x_2 \oplus x_3 \oplus x_1x_3 \oplus x_1x_4 \oplus x_0x_1x_4 \oplus x_2x_3x_4 \oplus x_0x_5 \oplus x_1x_3x_5 \oplus x_0x_4x_5 \\
&\quad \oplus x_1x_6 \oplus x_3x_6 \oplus x_0x_3x_6 \oplus x_5x_6 \oplus 1 \\
y_5 &= x_2 \oplus x_0x_2 \oplus x_0x_3 \oplus x_1x_2x_3 \oplus x_0x_2x_4 \oplus x_0x_5 \oplus x_2x_5 \oplus x_4x_5 \oplus x_1x_6 \oplus x_1x_2x_6 \\
&\quad \oplus x_0x_3x_6 \oplus x_3x_4x_6 \oplus x_2x_5x_6 \oplus 1 \\
y_6 &= x_1x_2 \oplus x_0x_1x_3 \oplus x_0x_4 \oplus x_1x_5 \oplus x_3x_5 \oplus x_6 \oplus x_0x_1x_6 \oplus x_2x_3x_6 \oplus x_1x_4x_6 \\
&\quad \oplus x_0x_5x_6
\end{aligned}
$$

In the presentation of the look-up table for S7 the input and output are represented using decimal numbers from 0 to 127. The look-up table of S7 is as follows:

54, 50, 62, 56, 22, 34, 94, 96, 38, 6, 63, 93, 2, 18, 123, 33
55, 113, 39, 114, 21, 67, 65, 12, 47, 73, 46, 27, 25, 111, 124, 81
53, 9, 121, 79, 52, 60, 58, 48, 101, 127, 40, 120, 104, 70, 71, 43
20, 122, 72, 61, 23, 109, 13, 100, 77, 1, 16, 7, 82, 10, 105, 98
117, 116, 76, 11, 89, 106, 0, 125, 118, 99, 86, 69, 30, 57, 126, 87
112, 51, 17, 5, 95, 14, 90, 84, 91, 8, 35, 103, 32, 97, 28, 66
102, 31, 26, 45, 75, 4, 85, 92, 37, 74, 80, 49, 68, 29, 115, 44
64, 107, 108, 24, 110, 83, 36, 78, 42, 19, 15, 41, 88, 119, 59, 3

Example The following is given in the specification to illustrate the relationship between the two representations. Given an input value 38, the value found in the S7 look-up table at position 38 is 58. Hence S7[38] = 58. Using the gate logic representation the same result is obtained as follows. First the input 38 is converted to a bit string:

$$38 = 0100110_2$$

from this, recalling that x_0 is the least significant bit, we derive that:

$$x_6 = 0, \quad x_5 = 1, \quad x_4 = 0, \quad x_3 = 0, \quad x_2 = 1, \quad x_1 = 1, \quad x_0 = 0$$

Substituting this input value for the gate logic equations gives us the following output:

$$\begin{aligned}
y_0 &= 0 \oplus 0 \oplus 0 \oplus 1 \oplus 1 \oplus 0 \oplus 0 \oplus 0 \oplus 0 \oplus 0 \oplus 0 \oplus 0 \oplus 0 &&= 0\\
y_1 &= 0 \oplus 0 \oplus 0 \oplus 1 \oplus 1 \oplus 0 \oplus 0 \oplus 0 \oplus 0 \oplus 0 \oplus 1 &&= 1\\
y_2 &= 0 \oplus 0 \oplus 0 \oplus 0 \oplus 0 \oplus 1 \oplus 0 \oplus 0 \oplus 0 \oplus 0 \oplus 0 \oplus 1 &&= 0\\
y_3 &= 1 \oplus 0 \oplus 0 \oplus 0 \oplus 0 \oplus 0 \oplus 0 \oplus 0 \oplus 0 \oplus 0 &&= 1\\
y_4 &= 0 \oplus 0 \oplus 0 \oplus 0 \oplus 0 \oplus 0 \oplus 0 \oplus 0 \oplus 0 \oplus 0 \oplus 0 \oplus 0 \oplus 0 \oplus 1 &&= 1\\
y_5 &= 1 \oplus 0 \oplus 0 \oplus 0 \oplus 0 \oplus 0 \oplus 1 \oplus 0 \oplus 0 \oplus 0 \oplus 0 \oplus 0 \oplus 0 \oplus 1 &&= 1\\
y_6 &= 1 \oplus 0 \oplus 0 \oplus 1 \oplus 0 \oplus 0 \oplus 0 \oplus 0 \oplus 0 \oplus 0 &&= 0
\end{aligned}$$

Thus $y = 0111010_2 = 58$.

The gate logic equations for S-box S9 are as follows:

$$y_0 = x_0x_2 \oplus x_3 \oplus x_2x_5 \oplus x_5x_6 \oplus x_0x_7 \oplus x_1x_7 \oplus x_2x_7 \oplus x_4x_8 \oplus x_5x_8 \oplus x_7x_8 \oplus 1$$

$$y_1 = x_1 \oplus x_0x_1 \oplus x_2x_3 \oplus x_0x_4 \oplus x_1x_4 \oplus x_0x_5 \oplus x_3x_5 \oplus x_6 \oplus x_1x_7 \oplus x_2x_7 \oplus x_5x_8 \oplus 1$$

$$y_2 = x_1 \oplus x_0x_3 \oplus x_3x_4 \oplus x_0x_5 \oplus x_2x_6 \oplus x_3x_6 \oplus x_5x_6 \oplus x_4x_7 \oplus x_5x_7 \oplus x_6x_7 \oplus x_8 \oplus x_0x_8 \oplus 1$$

$$y_3 = x_0 \oplus x_1x_2 \oplus x_0x_3 \oplus x_2x_4 \oplus x_5 \oplus x_0x_6 \oplus x_1x_6 \oplus x_4x_7 \oplus x_0x_8 \oplus x_1x_8 \oplus x_7x_8$$

$$y_4 = x_0x_1 \oplus x_1x_3 \oplus x_4 \oplus x_0x_5 \oplus x_3x_6 \oplus x_0x_7 \oplus \oplus x_6x_7 \oplus x_1x_8 \oplus x_2x_8 \oplus x_3x_8$$

$$y_5 = x_2 \oplus x_1x_4 \oplus x_4x_5 \oplus x_0x_6 \oplus x_1x_6 \oplus x_3x_7 \oplus x_4x_7 \oplus x_6x_7 \oplus x_5x_8 \oplus x_6x_8 \oplus x_7x_8 \oplus 1$$

$$y_6 = x_0 \oplus x_2x_3 \oplus x_1x_5 \oplus x_2x_5 \oplus x_4x_5 \oplus x_3x_6 \oplus x_4x_6 \oplus x_5x_6 \oplus x_7 \oplus x_1x_8 \oplus x_3x_8 \oplus x_5x_8 \oplus x_7x_8$$

$$y_7 = x_0x_1 \oplus x_0x_2 \oplus x_1x_2 \oplus x_3 \oplus x_0x_3 \oplus x_2x_3 \oplus x_4x_5 \oplus x_2x_6 \oplus x_3x_6 \oplus x_2x_7 \oplus x_5x_7 \oplus x_8 \oplus 1$$

$$y_8 = x_0x_1 \oplus x_2 \oplus x_1x_2 \oplus x_3x_4 \oplus x_1x_5 \oplus x_2x_5 \oplus x_1x_6 \oplus x_4x_6 \oplus x_7 \oplus x_2x_8 \oplus x_3x_8$$

In the presentation of the look-up table for S9 the input and output are represented using decimal numbers from 0 to 255. The look-up table of S9 is as follows:

167, 239, 161, 379, 391, 334, 9, 338, 38, 226, 48, 358, 452, 385, 90, 397
183, 253, 147, 331, 415, 340, 51, 362, 306, 500, 262, 82, 216, 159, 356, 177
175, 241, 489, 37, 206, 17, 0, 333, 44, 254, 378, 58, 143, 220, 81, 400
95, 3, 315, 245, 54, 235, 218, 405, 472, 264, 172, 494, 371, 290, 399, 76
165, 197, 395, 121, 257, 480, 423, 212, 240, 28, 462, 176, 406, 507, 288, 223
501, 407, 249, 265, 89, 186, 221, 428, 164, 74, 440, 196, 458, 421, 350, 163
232, 158, 134, 354, 13, 250, 491, 142, 191, 69, 193, 425, 152, 227, 366, 135
344, 300, 276, 242, 437, 320, 113, 278, 11, 243, 87, 317, 36, 93, 496, 27
487, 446, 482, 41, 68, 156, 457, 131, 326, 403, 339, 20, 39, 115, 442, 124
475, 384, 508, 53, 112, 170, 479, 151, 126, 169, 73, 268, 279, 321, 168, 364
363, 292, 46, 499, 393, 327, 324, 24, 456, 267, 157, 460, 488, 426, 309, 229
439, 506, 208, 271, 349, 401, 434, 236, 16, 209, 359, 52, 56, 120, 199, 277
465, 416, 252, 287, 246, 6, 83, 305, 420, 345, 153, 502, 65, 61, 244, 282
173, 222, 418, 67, 386, 368, 261, 101, 476, 291, 195, 430, 49, 79, 166, 330
280, 383, 373, 128, 382, 408, 155, 495, 367, 388, 274, 107, 459, 417, 62, 454
132, 225, 203, 316, 234, 14, 301, 91, 503, 286, 424, 211, 347, 307, 140, 374
35, 103, 125, 427, 19, 214, 453, 146, 498, 314, 444, 230, 256, 329, 198, 285
50, 116, 78, 410, 10, 205, 510, 171, 231, 45, 139, 467, 29, 86, 505, 32
72, 26, 342, 150, 313, 490, 431, 238, 411, 325, 149, 473, 40, 119, 174, 355
185, 233, 389, 71, 448, 273, 372, 55, 110, 178, 322, 12, 469, 392, 369, 190
1, 109, 375, 137, 181, 88, 75, 308, 260, 484, 98, 272, 370, 275, 412, 111
336, 318, 4, 504, 492, 259, 304, 77, 337, 435, 21, 357, 303, 332, 483, 18
47, 85, 25, 497, 474, 289, 100, 269, 296, 478, 270, 106, 31, 104, 433, 84
414, 486, 394, 96, 99, 154, 511, 148, 413, 361, 409, 255, 162, 215, 302, 201
266, 351, 343, 144, 441, 365, 108, 298, 251, 34, 182, 509, 138, 210, 335, 133
311, 352, 328, 141, 396, 346, 123, 319, 450, 281, 429, 228, 443, 481, 92, 404
485, 422, 248, 297, 23, 213, 130, 466, 22, 217, 283, 70, 294, 360, 419, 127
312, 377, 7, 468, 194, 2, 117, 295, 463, 258, 224, 447, 247, 187, 80, 398
284, 353, 105, 390, 299, 471, 470, 184, 57, 200, 348, 63, 204, 188, 33, 451
97, 30, 310, 219, 94, 160, 129, 493, 64, 179, 263, 102, 189, 207, 114, 402
438, 477, 387, 122, 192, 42, 381, 5, 145, 118, 180, 449, 293, 323, 136, 380
43, 66, 60, 455, 341, 445, 202, 432, 8, 237, 15, 376, 436, 464, 59, 461

Example The following is given in the specification to illustrate the relationship between the two representations. Given an input value 138, the value found in the S9 look-up table at position 138 is 339. Hence S9[138] = 339. Using the gate logic representation the same result is obtained as follows. First the input 138 is converted to a bit string:

$$138 = 010001010_2$$

from this, recalling once again that x_0 is the least significant bit, we derive that:

$$x_8 = 0, \quad x_7 = 1 \quad x_6 = 0, \quad x_5 = 0, \quad x_4 = 0, \quad x_3 = 1, \quad x_2 = 0, \quad x_1 = 1, \quad x_0 = 0$$

Substituting this input value for the gate logic equations gives us the following output:

$$\begin{aligned}
y_0 &= 0 \oplus 1 \oplus 0 \oplus 0 \oplus 0 \oplus 1 \oplus 0 \oplus 0 \oplus 0 \oplus 0 \oplus 1 && = 1 \\
y_1 &= 1 \oplus 0 \oplus 0 \oplus 0 \oplus 0 \oplus 0 \oplus 0 \oplus 0 \oplus 1 \oplus 0 \oplus 0 \oplus 1 && = 1 \\
y_2 &= 1 \oplus 0 \oplus 0 \oplus 0 \oplus 0 \oplus 0 \oplus 0 \oplus 0 \oplus 0 \oplus 0 \oplus 0 \oplus 1 && = 0 \\
y_3 &= 0 \oplus 0 \oplus 0 \oplus 0 \oplus 0 \oplus 0 \oplus 0 \oplus 0 \oplus 0 \oplus 0 && = 0 \\
y_4 &= 0 \oplus 1 \oplus 0 \oplus 0 \oplus 0 \oplus 0 \oplus 0 \oplus 0 \oplus 0 && = 1 \\
y_5 &= 0 \oplus 0 \oplus 0 \oplus 0 \oplus 0 \oplus 1 \oplus 0 \oplus 0 \oplus 0 \oplus 0 \oplus 0 \oplus 1 && = 0 \\
y_6 &= 0 \oplus 0 \oplus 0 \oplus 0 \oplus 0 \oplus 0 \oplus 0 \oplus 0 \oplus 1 \oplus 0 \oplus 0 \oplus 0 \oplus 0 && = 1 \\
y_7 &= 0 \oplus 0 \oplus 0 \oplus 1 \oplus 0 \oplus 0 \oplus 0 \oplus 0 \oplus 0 \oplus 0 \oplus 0 \oplus 0 \oplus 1 && = 0 \\
y_8 &= 0 \oplus 0 \oplus 0 \oplus 0 \oplus 0 \oplus 0 \oplus 0 \oplus 0 \oplus 1 \oplus 0 \oplus 0 && = 1
\end{aligned}$$

Thus $y = 101010011_2 = 339$.

7.4.3 *Key schedule*

KASUMI has a 128-bit key K. Each round of KASUMI uses 128 bits of key that are derived from K. Before the round keys can be calculated two arrays of 16-bit values K_j and K'_j $(j = 1, \ldots, 8)$ are derived in the following manner. The first array K_1, $K_2, \ldots, K_8$ is derived by subdivision of K into eight 16-bit sub-blocks such that:

$$K = K_1 \parallel K_2 \parallel K_3 \parallel K_4 \parallel K_5 \parallel K_6 \parallel K_7 \parallel K_8$$

The second array $K'_1, K'_2, \ldots, K'_8$ is derived from the first array by adding an array of 16-bit constants C_j as follows:

$$K'_j = K_j \oplus C_j$$

where the constants C_j are as given in Table 7.1. Then the subkeys (KL, KO and KI) are derived as defined by the following Table 7.2 using cyclic shift, where the following notation is used:

$D \lll n$ the left circular rotation of a datum D by n bits

Specifically, $D \lll 1 = \mathrm{ROL}(D)$, using the notation defined in Section 7.4.2.3.

Table 7.1 Constants C_j

C_1	0x0123
C_2	0x4567
C_3	0x89AB
C_4	0xCDEF
C_5	0xFEDC
C_6	0xBA98
C_7	0x7654
C_8	0x3210

Table 7.2 Definition of subkeys in KASUMI

	1	2	3	4	5	6	7	8
$KL_{i,1}$	$K_1 \lll 1$	$K_2 \lll 1$	$K_3 \lll 1$	$K_4 \lll 1$	$K_5 \lll 1$	$K_6 \lll 1$	$K_7 \lll 1$	$K_8 \lll 1$
$KL_{i,2}$	K'_3	K'_4	K'_5	K'_6	K'_7	K'_8	K'_1	K'_2
$KO_{i,1}$	$K_2 \lll 5$	$K_3 \lll 5$	$K_4 \lll 5$	$K_5 \lll 5$	$K_6 \lll 5$	$K_7 \lll 5$	$K_8 \lll 5$	$K_1 \lll 5$
$KO_{i,2}$	$K_6 \lll 8$	$K_7 \lll 8$	$K_8 \lll 8$	$K_1 \lll 8$	$K_2 \lll 8$	$K_3 \lll 8$	$K_4 \lll 8$	$K_5 \lll 8$
$KO_{i,3}$	$K_7 \lll 13$	$K_8 \lll 13$	$K_1 \lll 13$	$K_2 \lll 13$	$K_3 \lll 13$	$K_4 \lll 13$	$K_5 \lll 13$	$K_6 \lll 13$
$KI_{i,1}$	K'_5	K'_6	K'_7	K'_8	K'_1	K'_2	K'_3	K'_4
$KI_{i,2}$	K'_4	K'_5	K'_6	K'_7	K'_8	K'_1	K'_2	K'_3
$KI_{i,3}$	K'_8	K'_1	K'_2	K'_3	K'_4	K'_5	K'_6	K'_7

7.5 Mathematical Analysis of KASUMI by the Task Force

7.5.1 Properties of components

Each functional component of KASUMI was examined during the design and evaluation process performed by the Task Force. The mathematical properties were identified and analysed to see if any of the known mathematical structures caused any weakness that could be used as a basis for an attack on the entire algorithm. A brief overview of the results of this work is given in this section. For the complete report by the Task Force, see [17].

7.5.1.1 FL function

The FL function is a linear function, but the security of the algorithm is not meant to depend on it. Its main purpose is to be a low-cost additional scrambling, making individual bits harder to track through the rounds. The FL function has the property that for any key KL, an input of $0^{16}1^{16}$ always gives an output of 1^{32}. Hence for some round inputs, some of the key bits in KL can be changed without having any effect on the output of that round. This property can be used to guarantee a zero difference at the end of the first round, thus effectively removing the first round. More generally, small changes to the input of FL only make small output changes, and this can be useful going either forward or backward through the FL.

The fixed point is used in some of the differential attacks mentioned later, but no attack exploiting this property that extends beyond five rounds of KASUMI has been found.

7.5.1.2 FI function

This is the basic randomizing function of KASUMI with 16-bit input and 16-bit output. It is again composed of a four-round structure using two nonlinear substitution boxes S7 and S9. By theorem 4 of [83], the average linear and differential probability of FI is less than $(2^{-9+1})(2^{-7+1}) = 2^{-14}$, assuming uniform distribution of the subkeys in use. The S-boxes S7 and S9 have been designed to avoid linear structures in FI. This fact has also been confirmed by statistical testing.

The Walsh spectra of the outputs of FI for several keys have also been computed. They behave as expected for such small functions with considerably low algebraic degree.

7.5.1.3 FO function

The FO function constitutes the nonlinear part of the KASUMI round function. Again using theorem 4 of [83], it can be seen that the average linear and differential probability of FO is less than 2^{-28}, assuming uniform distribution of the subkeys in use. For any fixed key, FO is a permutation of 32-bit blocks, but due to its three-round structure it can be distinguished from a randomly chosen permutation using four chosen plaintexts.

Consideration was given to improving the diffusion properties of the FO by adding a fourth round, as was done for the FI function. On the cost of adding complexity and power consumption this could improve to the general security margins of KASUMI. However, there are no indications that the properties of a three-round FO can be used in an attack on the full eight-round KASUMI.

7.5.1.4 The S7 box

The S7 box in KASUMI is essentially the same as S7 in MISTY1 [83]. Only the bit order before and after the original S7 has been rearranged. The S7 box is specially designed to be easy to implement in hardware using combinational logic, and the nonlinear order is 3. The algebraic normal form of this function is given in Section 7.4.2.6 and some of the properties of this box are described below.

Kasami exponent

The substitution box S7 is a linear transform of the monomial x^{81} defined over $GF(2^7)$, which has optimal nonlinearity properties [50]. The exponent 81 belongs to the set of so-called Kasami exponents. These are of the form $d = 2^{2k} - 2^k + 1 \pmod{2^n - 1}$, for $n = 2m + 1$, $2 \leq k \leq m$ and $\gcd(k, m) = 1$. If d is a Kasami exponent, then the power function x^d is maximally nonlinear. Exponent 81 cannot directly be given in Kasami form, but is equivalent to such an exponent. Indeed $81 = 2^6 + 2^4 + 1 = 2^4(2^4 - 2^2 + 1) \pmod{2^7 - 1}$, and for $n = 7$, $13 = 2^4 - 2^2 + 1$ is a Kasami exponent with $k = 2$. Since squaring (raising an element to power 2) is a linear operation in $GF(2^7)$, functions x^{13} and x^{81} have the same nonlinearity properties.

Probabilistic approximation

In probabilistic cryptanalysis we try to approximate a functional block using low-degree functions. Attacks based on this approach have been shown to be able to break block ciphers that were proven to be resistant against linear/differential attacks [64]. In the case of S7 it is especially interesting to see if the low-degree expression over GF(2) gives rise to probabilistic approximations of low degree over $GF(2^7)$. For this purpose, Sudan's algorithm, one of the newest advances in coding theory, was applied. For several trials, no significant approximation of degree 6 or lower was found. Thus it seems impossible to get a good probabilistic approximation of S7 over $GF(2^7)$.

Cycle structure

The cycle structure of the S7 permutation has been determined (Table 7.3): it has one fixed point given by S7(27) = 27, but no obvious deficiencies can be found from the cycle structure.

7.5.1.5 The S9 box

The S9 box is different from the S9 box in MISTY1, but was constructed in much the same way. It is easy to implement in hardware (actually easier than the original S9) and has the nonlinear order 2. The algebraic normal form of this function is given in Section 7.4.2.6. S9 can be seen as a composition of the power function x^5 and a linear

Table 7.3 The cycle structure of S7

Cycle length	No. of cycles
92	1
22	1
13	1
1	1

output transformation defined over GF(2^9). It is known to achieve almost perfect nonlinearity [93].

Linear structures

Since the component functions of S9 are quadratic, they are bound to have linear structures. A Boolean function f is said to have a linear structure if there is a vector w such that $f(x) \oplus f(x \oplus w)$ is constant as x varies. Each of the nine component functions of S9 has exactly one linear structure. Moreover, each linear combination of the components of S9 has exactly one linear structure [95].

The large number of linear structures in S9 could easily result in a linear structure over the entire FI function. However, this was taken into account in the selection of S9. The input and output bits of S9 were permuted in such a way that the unavoidable linear structures over S9 do not lead to any linear structures in the components of the FI and FO functions.

Cycle structure

The cycle structure of the S9 permutation has been determined (Table 7.4) and does not deviate from the expected structure for a random permutation.

Table 7.4 The cycle structure of S9

Cycle length	No. of cycles
275	1
121	1
74	1
26	1
12	1
2	1
1	2

7.5.1.6 Key schedule

The key schedule of KASUMI (see Section 7.4.3) is very simple, but this fact has not been found to constitute any real weakness, and there seems to be no gain in practice by making it more complicated. Each of the 128 bits of the secret key is used once and only once in every round. They are used in different ways in different rounds, at different parts within those rounds and at times the values are altered using key modification constants.

Due to the use of the constants C_1 to C_8 in the key schedule, there is no fixed recurrence relation between consecutive round keys. This property is required to prevent chosen plaintext attacks that are faster than an exhaustive search. Further, there exists no equivalent, more compact representation of the expanded key.

Even if regularity and symmetry in key scheduling do not introduce weaknesses in the algorithm, care should be taken that shorter keys, say 64 bit, are not extended to a full-length key in a very symmetric way. Just padding with 0's could give some advantage to an attacker (see Section 7.5.2.2) and should not be recommended.

In [83] Matsui showed that if subkey bits are independent, the average differential and linear probabilities are less than 2^{-56}. Some concern has been expressed that with the simple key schedule in KASUMI, the assumption of subkey independence might be too optimistic. However, no indications in this direction have been observed (see also Section 7.5.4).

7.5.2 Differential cryptanalysis

The KASUMI cipher was constructed in such a manner that, provided the subkeys are independent, three rounds of KASUMI have no differential or linear characteristics with probability larger than 2^{-56}. It should be noted that the theoretical upper bound for characteristics over FI is tight. It is possible to find differential characteristics for FI with probability 2^{-14}. It is also important to note that the differential effect of FL is low.

Differential attacks on KASUMI are very similar to those found for MISTY1. The distinguishing factor between the two ciphers is that FL functions are placed differently. Therefore all attacks known for MISTY1 without FL functions can also be applied to KASUMI without FL functions. Since in KASUMI the FL functions form a part of the round function, it follows that some of the attacks that apply to MISTY only if FL functions have been removed may be relevant for KASUMI even if FL functions are in place.

In this section the most important differential attacks on KASUMI are reviewed. All of them were known to the design and evaluation teams and considered to pose no real threat to KASUMI. It is interesting to note that since the publication of KASUMI in early 2000 no essentially new approaches to attack KASUMI have been presented.

7.5.2.1 A differential chosen plaintext attack

This section describes a chosen plaintext attack on five rounds of KASUMI, which can be used to recover the key. The attack requires roughly 2^{38} chosen plaintexts and 2^{80} small operations.

Let X_i ($i = 0, \ldots, 9$) denote the 32-bit word that is the right data half taken as output from round i ($i = 0, \ldots, 8$) or, equivalently, the left data half taken as input to round i ($i = 1, \ldots, 9$). Thus the plaintext is $[X_1, X_0]$ and the ciphertext is $[X_9, X_8]$. The operation at round i is summarized by the relation $X_{i+1} = f_i(X_i) \oplus X_{i-1}$ ($i = 1, \ldots, 8$), where $f_i = \mathrm{FO}_i \circ \mathrm{FL}_i$ (if i is odd) and $f_i = \mathrm{FL}_i \circ \mathrm{FO}_i$ (if i is even) (see Section 7.4.2.2). Then $X_4 \oplus X_0$ is a one-to-one function of X_0. This means that when performing 2^{32} encryptions with the same X_1 but different right plaintext halves X_0, no collision will occur among the obtained $X_4 \oplus X_0$ values. It follows that for a fixed X_1 the XOR of all $X_4 \oplus X_0$ blocks, as X_0 varies over all possible 2^{32} values, is equal to 0. Since the XOR of all such X_0 blocks is equal to 0, it follows that the XOR over all such X_4 blocks is equal to 0. The attack makes use of this property, which is formalized in [17] as follows:

(P) Given fixed X_1 and a fixed encryption key K denote by $\mathrm{SUM}(K, X_1)$ the XOR of all 2^{32} values X_4 (possibly not all different) associated with the 2^{32} possible X_0 values under encryption with KASUMI. Then $\mathrm{SUM}(K, X_1) = 0$ for all K and X_1.

The attack requires a number of chosen plaintexts (X_1, X_0), formed by a small number, say 2^6, of fixed, arbitrary values of X_1 and all the 2^{32} values of X_0. This makes a total of 2^{38} plaintexts. For each of the values X_1, the 2^{32} plaintexts (X_1, X_0), in which the left half equals X_1, are considered. Assume that the attacker has access to the corresponding ciphertexts (X_6, X_5) after the fifth round. The attack uses property (P) to test parts of the fifth round key. When the 82-bit $(\mathrm{KL}_5, \mathrm{KO}_{5,1}, \mathrm{KI}_{5,1,2}, \mathrm{KO}_{5,2}, \mathrm{KI}_{5,2,2})$ value has been found, the remaining 42 unknown bits of K can be determined by exhaustive search.

Using linearization methods similar to those described by Tanaka et al. in [133], it might be possible to extend this attack to six rounds, but not to the full eight rounds of KASUMI.

Other types of attacks considered by the evaluation team were based on differentials with low Hamming weight. Recall that FL functions have the property that an input difference with low Hamming weight evolves to an output difference with a low Hamming weight. However, the same property does not hold for FO. Therefore, no substantial attack on KASUMI could be found.

7.5.2.2 Differential related key attacks

Related key attacks seem to be of no threat in the 3GPP context. The confidentiality and integrity keys are derived in the AKA process, therefore making it impossible for an attacker to have any control over the keys.

For completeness, the Task Force also considered related key attacks for KASUMI. The conclusion was that it is possible to carry out differential related key attacks on four and five rounds of KASUMI. The four-round attack requires the encryptions of approximately 2^9 chosen plaintext pairs X and X^* under keys K and K^* respectively, where K and K^* differ in only one bit. The average complexity of this attack is approximately 2^{41}. The five-round attack, which is an extension of the four-round attack, requires the encryptions of on average 3×2^{17} chosen plaintext pairs and has an average complexity of approximately 2^{36}.

The requirement specification [10] states that if the key needs to be shorter than 128 bits, the least significant bits should be set to 0. If the key is reduced to only 64 bits (i.e., $K_5 = \cdots = K_8 = 0^{16}$) the algorithm is vulnerable to a five-round related key attack that only needs about 10 plaintexts encrypted under two keys and has a complexity of roughly 2^{25}. These attacks all rely on essentially the same differential, which predicts differences at the end of the third round with a probability of $\frac{1}{2}$. This differential arises since the subkey K_3 appears early in all of the first three rounds. Changing the order in which the subkeys are fed into the rounds, for example, could destroy this differential, but such a change could create other differentials in other rounds.

In any event, no attack of this type has been found that extends beyond five rounds of KASUMI and, as stated, related key attacks are no threat in the 3GPP context.

7.5.2.3 Impossible differentials

Impossible differentials are differentials with probability 0. Such a property can be used to distinguish the cipher from a truly random function or to test for a correct key.

In the FI function there are no impossible differentials, because of its four-round structure. In the three-round FO function, however, several impossible differentials occur since the round function FI is bijective. These lead to impossible differentials over two and three rounds of KASUMI without the FL function.

FL functions seem to destroy most of these impossible differentials or, more precisely, make their existence key-dependent. No impossible differentials are known to exist for the complete KASUMI that could be derived from the impossible differentials of the FO function. Hence the only known key-independent differential for the KASUMI cipher is the well-known five-round impossible differential of the form:

$$[0, A] \rightarrow [A, 0] \rightarrow [*, A] \rightarrow [A, *] \rightarrow [0, A] \rightarrow [A, 0]$$

where A is a nonzero 32-bit block, 0 is a 32-bit block made up of 0's and each occurrence of $*$ can be replaced by any (possibly different) nonzero blocks of 32 bits. All Feistel ciphers with bijective round functions have the property that this kind of differential can never occur [68]. Note that it is not applicable to MISTY1 due to differently placed FL functions.

Recall now the notation for X_i used in Section 7.5.2.1 and let ΔX_i denote the XOR difference of two data block values for X_i. In the above expression of the impossible differential we have $A = \Delta X_0 = \Delta X_2 = \Delta X_4 = \Delta X_6$. Hence this impossible differential over five rounds, as observed from the input differences to the first round and from the output differences from the last round, can be expressed as follows:

$$\text{It is impossible that} \quad \Delta X_1 = \Delta X_5 = 0 \quad \text{and} \quad \Delta X_6 = \Delta X_0 \neq 0$$

It follows that KASUMI restricted to five rounds can be distinguished from a random permutation with a probability close to 1 with slightly more than 2^{32} chosen plaintext/ciphertext pairs.

The external evaluation teams developed more complex attacks based on this impossible differential to discover part of the key. One such attack works on KASUMI that is reduced to six rounds. It requires 2^{55} chosen plaintexts and computation of approximately 2^{119} FI values. Another attack against six rounds of KASUMI was found to require $2^{53.3}$ chosen plaintexts with a complexity of the order of 2^{100} encryptions. Both attacks exploit impossible differentials and the structure of the FO function. This result is similar to the attack independently discovered by Ulrich Kühn [74], which is described in Section 7.6.

No similar attacks on the full eight rounds of KASUMI have been found, and in the 3GPP context these attacks are not applicable.

7.5.3 Truncated differentials

Truncated differentials are generalizations of differentials. Instead of fixing a difference it is allowed to vary in a relatively small set. Truncated differentials were also studied for KASUMI. The best way that has been found to exploit truncated differentials for KASUMI leads to an attack on three or four rounds of KASUMI without the FL function. This attack uses the fact that the FO function that is restricted to the 16 leftmost input bits is bijective on the leftmost 16 bits in the output. Three rounds can be broken using about 2^{35} plaintext pairs derived from 2^{18} chosen plaintexts. The four-round attack requires 2^{48} chosen plaintexts. The FL function will complicate the attack, and, in any event, KASUMI with five rounds or more is secure against this attack.

7.5.4 Linear cryptanalysis

The validity of the proofs of security by Matsui [83] was also examined. The question is: how average is the behaviour of fixed keys with respect to linear approximations over the FI function? Mathematical calculations using the Walsh–Hadamard transform and experimental calculations were carried out independently and reached the same conclusions.

For most key values, linear probabilities were found to be smaller than the theoretical upper bound 2^{-14} derived by Matsui, but there are specific key values and linear hulls for the FI function with linear probabilities about 2^{-12}. Of course, there are also key values for which actual linear predictability is much lower than the average case. But even the maximal correlations that were found in this analysis are not high enough to make it possible to chain them to a useful linear approximation path over several rounds of KASUMI.

It is mentioned in the evaluation report that one attack in five rounds of KASUMI was found, which would require a work effort of at least 2^{95} operations and around 2^{58} known plaintexts. But this attack would be applicable only to a fraction of 2^{-3} of the key space. A variant may potentially reduce the work effort to 2^{93} and require around 2^{49} known plaintexts, but will only be applicable to a fraction of 2^{-41} of the key space. Details of these attacks are neither given in [17] nor published elsewhere.

The conclusion drawn from the analysis was that all keys of the FI function behave pretty much like an average key with respect to known linear approximation relations.

7.5.5 Higher order differential attacks

Most of the cryptanalysis on MISTY1 has concentrated on higher order differential cryptanalysis. The differential property leading to this attack is actually due to the choice of S7 box. While KASUMI essentially uses the same S7 function, the effect of the differential property has been eliminated by adding a fourth round in the FI function.

Therefore, no indication has been found that the seventh-order differential property known to MISTY1 still holds for KASUMI. The design and evaluation teams of KASUMI believe that traditional attacks based on higher order differentials will work for at most five rounds of KASUMI. No other variants have been found that work for more than five rounds of KASUMI.

7.6 Public Research on KASUMI

After publication of KASUMI, public scrutiny of MISTY-type algorithms also extended to KASUMI algorithms. Cryptographers apply different analytical methods in their efforts to find defects in these algorithms. So far, the algorithms have resisted the attacks well and remain unbroken.

The method using higher order differentials had some further developments. Babbage and Frisch showed that such higher order differentials occurred even if the selected S7 transformation was replaced by some other highly nonlinear function of algebraic degree 3 [37]. In a subsequent study Canteaut and Videau [45] analysed this phenomenon. The question concerns the speed at which the degree of composed functions increases. Typically, after two applications of functions with degree 3 components we have a function of degree 9, three subsequent applications would produce a function with components of degree 27 and so on. It was shown in [45] that optimally nonlinear functions have the following property. When composed with some other function the algebraic degree grows at a significantly slower pace. The S7 and S9 functions of MISTY1 are typical examples of such functions. The S-box selected for AES is also based on a highly nonlinear power function x^{-1} [90]. Its nonlinearity is high but not the highest possible, which is the case for S7 and S9. Moreover, it does not have the property that causes the algebraic degree of composed functions to increase slowly.

At the Eurocrypt 2001 conference, Ulrich Kühn [74] presented his analysis of reduced-round MISTY and applied it to KASUMI. His attack was based on a five-round impossible differential, which is known to exist for a Feistel structure with bijective round functions. These results were already known to the designer team and reported in [17] (see also Section 7.5.2.3). Kühn's main result on MISTY1 in [74] was a method that could find part of the key after six rounds of MISTY1 without any FL functions. This attack would require 2^{54} chosen plaintexts and a computation equivalent to 2^{61} encryptions using the six-round MISTY1 without FL functions. It was not possible to attack MISTY1 with FL functions using impossible differentials. Later, in [75], Kühn applied a special technique called the "slicing attack" to analyse four rounds of MISTY1 with FL functions with a complexity of $2^{22.25}$ of data and 2^{45} of computation time. At the Fast Software Encryption (FSE) 2002 workshop, Knudsen and Wagner applied a different analytical technique, "integral cryptanalysis", to obtain a key recovery attack on five rounds of MISTY1 with FL functions with a complexity of 2^{34} of data and 2^{48} of computation time [71].

MISTY1 and KASUMI constructions have also been proven to provide pseudorandomness. The early paper by Sakurai and Zheng [128] had shown that

MISTY structures were not as efficient at providing pseudorandomness as the Feistel construction used by the DES algorithm. Later, however, Gilbert and Minier [54] and Kang et al. [66] showed independently that four-round, MISTY-type transformations are pseudorandom permutations. Later, Kang et al. also provided a proof of the security of the KASUMI construction in [67] by showing that the four-round KASUMI is indistinguishable from a random permutation.

One possible line of attacks that may still occur involves those algebraic methods that are already applied to the AES block cipher [47], [85], see also Section 8.9.6. The S-boxes of KASUMI are based on algebraic, low-degree power functions that are similar to the S-boxes of the AES. While such attacks, if they happen at all, are not expected to cause any security threat to practical 3GPP applications, 3GPP has already initiated work to develop a new fallback cipher in case of serious failure of the f8 algorithm based on KASUMI.

7.7 Implementation Issues

7.7.1 Parallel operation

KASUMI has been designed so that it can be efficiently implemented in hardware. Let us now highlight the most important properties that can be exploited in such a hardware implementation. Because the list given in the specification is not exhaustive each implementor is given the opportunity to design his or her own optimizations. First, a simple key schedule is easy to implement in hardware. Second, the S7 and S9 substitution boxes have been designed in such a way that they may be implemented using a little combinational logic rather than by using large look-up tables.

One of the main reasons Matsui introduced MISTY-type, new cipher structures for FI and FO functions, instead on relying on the traditional and well-known Feistel structure, was that such structures allow two consecutive round functions to be implemented in parallel, thus increasing the speed of encryption. As a result KASUMI has the following two optimization options:

- operation of the S7 and S9 functions in the FI function can be carried out in parallel; and
- operation of the $FI_{i,1}$ and $FI_{i,2}$ functions in the FO function can be carried out in parallel.

These two options are depicted in Figure 7.4.

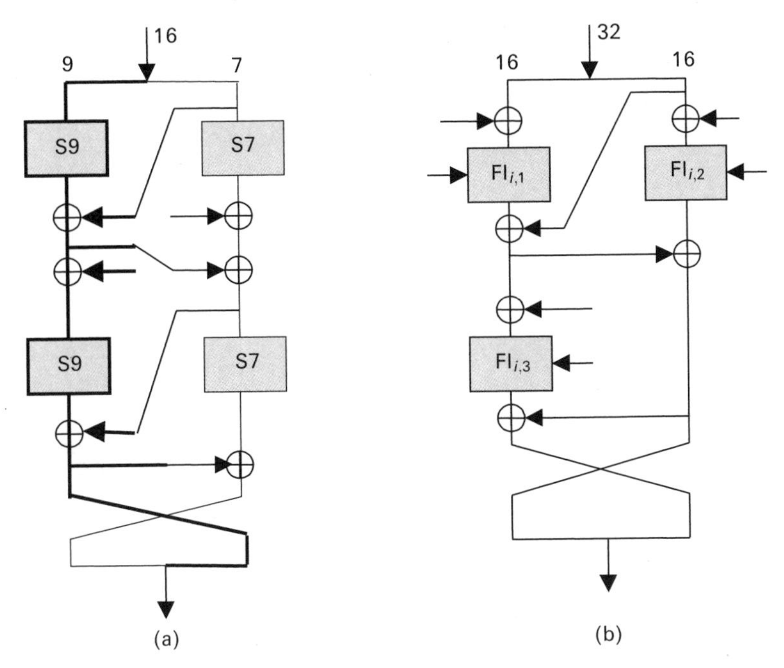

Figure 7.4 Optimization options in the FI and FO functions

7.7.2 Implementation attacks

KASUMI was also analysed by the Task Force with respect to such differential attacks as timing attacks, simple power analysis and differential power analysis. This investigation did not reveal any properties of KASUMI that would make it particularly vulnerable to these types of attacks. Specifically, KASUMI's key scheduling is particularly favourable against power attack methods that try to derive information about the Hamming weight of subkey bytes. The restricted use of KASUMI in the 3GPP environment will also reduce the possibilities of such attacks. In an application where an attacker can take measurements of time of execution and/or power consumption, special care should be taken to guarantee resistance against implementation attacks

8

Authentication and Key Generation Algorithm

8.1 Design Task Force

Authentication and key generation algorithms for UMTS need not be standardized. Operators are given the opportunity to choose the algorithm they want to rely on for network access and call origin authentication. Algorithms will be implemented at the network operator's Authentication Centres (AuCs) and in the USIMs (Universal Subscriber Identity Modules) of the mobile devices of the same operator's subscribers. Hence there is no need for interoperability between different operators. However, to achieve interoperability between different USIM implementations and the AuC version of the algorithm may require substantial effort, and it may well be easier to use a standard algorithm. Moreover, the design and implementation of a robust cryptographic algorithm is never a trivial task and may not be an option for all operators.

Therefore, SA3 wanted to provide such an algorithm for 3GPP that could be used by operators who do not wish to provide one of their own. In January 2000, SA3 sent a communication (S3-000089) to ETSI SAGE to enquire about the possibilities of SAGE designing an example AKA function for 3GPP. SAGE was asked to consider whether it would be willing and able to act as the design authority for such an algorithm. It was also asked to follow a similar procedure to that used for the development of the confidentiality and integrity algorithms for 3GPP.

At the same time the GSM Association Security Group was planning a new, more robust GSM authentication function to replace the already broken COMP128 algorithm, also known as COMP128-1, and its replacement COMP128-2. SA3 saw some advantages in combining these design efforts and in basing both algorithms on the same core design.

It was decided that the algorithm would be published as a 3GPP specification and SA3 estimated a design period of about six months. The Task Force was officially nominated in July 2000.

In addition to the regular SAGE members, the special Task Force had members from two mobile network and handset manufacturers. Since the authentication algorithm is typically implemented on a smart card, the USIM card in the mobile handset, it was considered important that the Task Force should also include representatives from smart card manufacturers.

The design of the example 3GPP AKA functions was completed in December 2000 after about six months' intensive work by the SAGE Task Force. External evaluators were not used this time, partly because of severe time constraints and partly because the design was based on the Rijndael block cipher algorithm, which was shortly to become the AES standard. Moreover, the specific block cipher mode of operation was an extension of a standard one.

8.2 Requirements

8.2.1 Authentication specification

The UMTS AKA system defined in clause 6.3 of [1] makes use of a number of different types of cryptographic algorithms to perform various security tasks. Altogether eight different functions are required, two of which (f5 and f5*) are optional and only needed if the synchronization number is to be concealed. Recall that the synchronization number allows different authentication instances to be related to each other and, in this manner, may reveal the identity of the user (see Section 2.1.1.2).

The eight cryptographic functions of the UMTS authentication procedure are the following:

f0 the random challenge generating function;

f1 the network authentication function;

f1* the resynchronization message authentication function;

f2 the user authentication function (AUTN);

f3 the Cipher Key (CK) derivation function;

f4 the Integrity Key (IK) derivation function;

f5 the Anonymity Key (AK) derivation function for normal operation;

f5* the AK derivation function for resynchronization.

Let us now give an overview of how these functions are used. All authentication functions with the exception of f0 are "keyed" cryptographic functions, which means that they are given as families of functions with the key K as parameter. Application of function fi with key K is denoted as fi_K.

8.2.1.1 Generation of quintets at the AuC

To generate a quintet (Figure 8.1), the Home Location Register (HLR)/AuC first generates a random number (RAND) using the f0 function. Then it computes a Message Authentication Code (MAC) for authentication MAC-$A = f1_K(SQN \| RAND \| AMF)$, an expected response $XRES = f2_K(RAND)$, a $CK = f3_K(RAND)$ and an $IK = f4_K(RAND)$. The HLR/AuC assembles the authentication token AUTN = SQN || AMF || MAC-A and forms the "quintet" Q = (RAND, XRES, CK, IK, AUTN). The quintet Q is sometimes called the "authentication vector" or "authentication quintuple".

If the SQN is to be concealed, the HLR/AuC also computes an $AK = f5_K(RAND)$ and computes the concealed SQN as $SQN \oplus AK$. In this case the AUTN in the quintet Q is formed as AUTN = $(SQN \oplus AK) \| AMF \| MAC\text{-}A$. Concealment of the SQN is optional.

8.2.1.2 Authentication and key derivation in the USIM

On receipt of a (RAND, AUTN) pair, the USIM acts as follows (Figure 3.16). First, it retrieves the unconcealed SQN. If the SQN is concealed, the USIM computes $AK = f5_K(RAND)$ and retrieves the SQN by computing $SQN = (SQN \oplus AK) \oplus AK$. Then the USIM computes XMAC-A = $f1_K(SQN \| RAND \| AMF)$, the user response $RES = f2_K(RAND)$, the $CK = f3_K(RAND)$ and the $IK = f4_K(RAND)$.

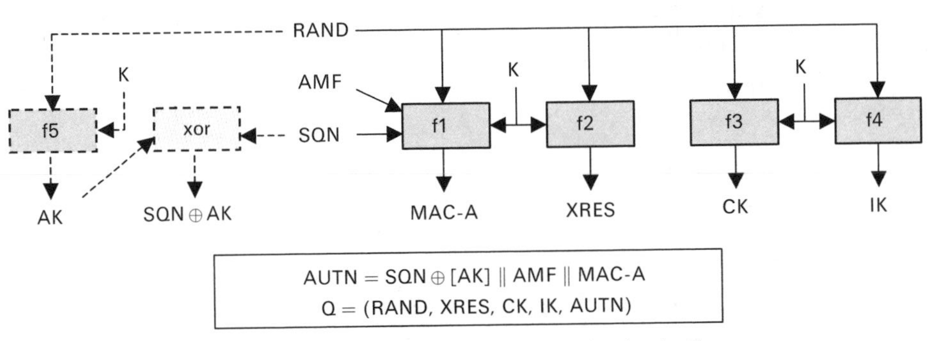

AUTN = SQN ⊕ [AK] || AMF || MAC-A
Q = (RAND, XRES, CK, IK, AUTN)

Figure 8.1 Generation of quintets in the AuC
AuC = Authentication Centre

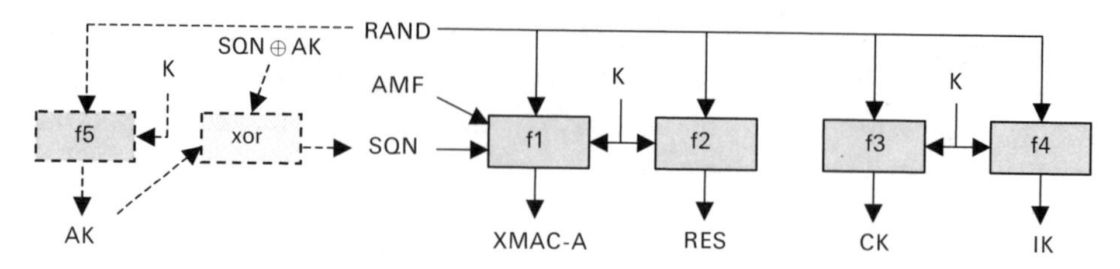

Figure 8.2 Authentication and key derivation in the USIM
USIM = Universal Subscriber Identity Module

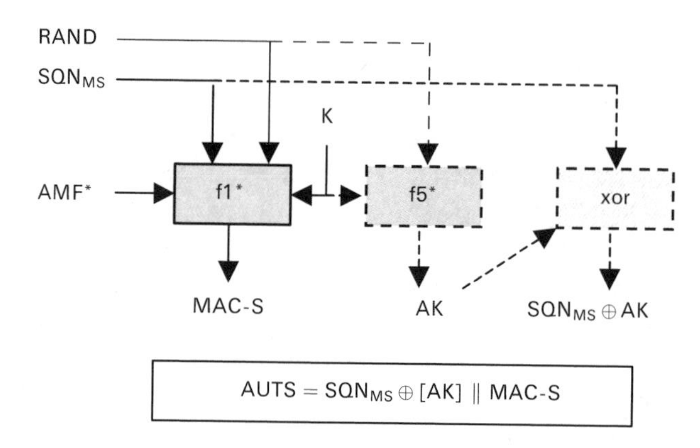

Figure 8.3 Generation of the resynchronization token in the USIM
USIM = Universal Subscriber Identity Module

8.2.1.3 Generation of the resynchronization token in the USIM

When there is a synchronization failure, the USIM generates a resynchronization token as follows (Figure 8.3). It computes MAC-S $= \mathrm{f1}^*_K(\mathrm{SQN}_{MS} \| \mathrm{RAND} \| \mathrm{AMF}^*)$, where AMF* is the default value for AMF used in resynchronization. Then, the resynchronization token is constructed as AUTS $= \mathrm{SQN}_{MS} \| \mathrm{MAC\text{-}S}$. If SQN_{MS} is to be concealed by means of an AK, the USIM computes AK $= \mathrm{f5}^*_K$ (RAND) and the concealed counter value is then computed as $\mathrm{SQN}_{MS} \oplus \mathrm{AK}$. In this case the resynchronization token is formed as AUTS $= (\mathrm{SQN}_{MS} \oplus \mathrm{AK}) \|$ MAC-S.

8.2.1.4 Resynchronization at the HLR/AuC

On receipt of an indication of synchronization failure and an (AUTS, RAND) pair, the HLR/AuC takes the following actions (Figure 8.4). It computes XMAC-S $=$ $\mathrm{f1}^*_K(\mathrm{SQN}_{MS} \| \mathrm{RAND} \| \mathrm{AMF}^*)$, where AMF* is the default value for AMF used

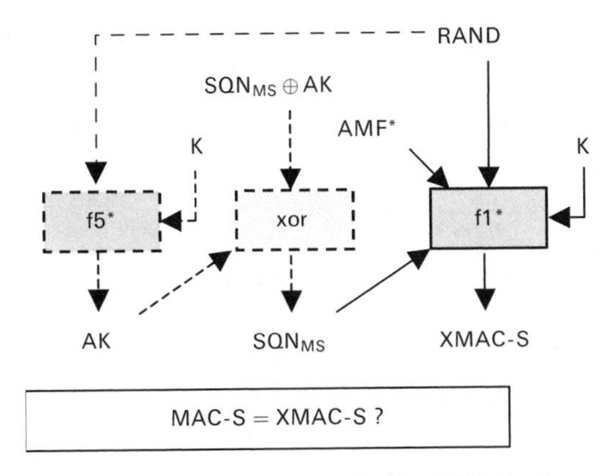

Figure 8.4 Resynchronization in the HLR/AuC
HLR = Home Location Register; AuC = Authentication Centre

in resynchronization. If SQN_{MS} is concealed by means of an AK, then the HLR/AuC must first compute $AK = f5^*_K(RAND)$ and retrieve the unconcealed counter value as $SQN_{MS} = (SQN_{MS} \oplus AK) \oplus AK$. Then, the AuC verifies whether MAC-S included in the AUTS is equal to the computed value XMAC-S.

8.2.2 Functional requirements for UMTS authentication

In document [10] the functional requirements for UMTS authentication are further specified. In this section we will describe them.

8.2.2.1 Use

Functions f0–f5 are only used to provide mutual entity authentication between the USIM and the AuC, to derive keys to protect user and signalling data transmitted over the radio access link and to conceal the SQN to protect user identity confidentiality. The f1* function is only used to provide data origin authentication for synchronization failure information sent by the USIM to the AuC. The f5* function is only used to provide user identity confidentiality during resynchronization.

8.2.2.2 Allocation

Functions f1–f5, f1* and f5* are allocated to the AuC and the USIM. The f0 function is allocated to the AuC.

8.2.2.3 Extent of standardization

Functions f0–f5, f1* and f5* are proprietary to the Home Environment (HE). Examples of f1, f1* and f2 functions are CBC MACs or HMACs [61].

8.2.2.4 Implementation and operational considerations

Functions f1–f5, f1* and f5* are designed so that they can be implemented on an Integrated Circuit (IC) card equipped with a 8-bit microprocessor running at 3 MHz with 8 kbyte of ROM and 300 byte of RAM and produce an AK, XMAC-A, RES, CK and IK in less than 500 ms of execution time.

8.2.2.5 Types of cryptographic functions

Apart from the intended use of the functions the following cryptographic requirements were allocated to each function in [10]:

- f0 is the random challenge-generating function. It should be a (pseudo) random number-generating function and map the internal state of the generator to the challenge value RAND.
- f1 is the network authentication function. It should be a keyed MAC function that takes the subscriber key K as the key and maps the data (SQN, RAND, AMF) to MAC-A (or XMAC-A). In particular, it must be computationally infeasible to derive K from knowledge of RAND, SQN, AMF and MAC-A (or XMAC-A).
- f1* is the resynchronization message authentication function. It should be a keyed MAC function that takes the subscriber key K as the key and maps the data (SQN, RAND, AMF*) to MAC-S (or XMAC-S). In particular, it must be computationally infeasible to derive K from knowledge of RAND, SQN, AMF* and MAC-S (or XMAC-S).
- f2 is the user authentication function. It should be a keyed MAC function that takes the subscriber key K as the key and maps the challenge RAND to RES (or XRES). In particular, it must be computationally infeasible to derive K from knowledge of RAND and RES (or XRES).
- f3 is the CK derivation function that takes the subscriber key K and the random challenge RAND as inputs and produces the CK as output. It should be a key derivation function. In particular, it must be computationally infeasible to derive K from knowledge of RAND and CK.

- f4 is the IK derivation function that takes the subscriber key K and the random challenge RAND as inputs and produces the IK as output. It should be a key derivation function. In particular, it must be computationally infeasible to derive K from knowledge of RAND and IK.

- f5 is the AK derivation function for normal operation that takes the subscriber key K and the random challenge RAND as inputs and produces the AK as output. It should be a key derivation function. In particular, it must be computationally infeasible to derive K from knowledge of RAND and AK. Its use is optional.

8.2.2.6 Interfaces of the authentication functions

In this section we define the lengths and types of input and output parameters of authentication functions. Note that the input (i.e., the random or pseudorandom internal state of function f0) remains unspecified.

Parameter K is the subscriber authentication key:

$$K[0], K[1], \ldots, K[127]$$

The length of K is 128 bits. It is a long-term secret key stored in the USIM and at the AuC.

Parameter RAND is the random challenge:

$$RAND[0], RAND[1], \ldots, RAND[127]$$

The length of RAND is 128 bits.

Parameter SQN is the sequence number:

$$SQN[0], SQN[1], \ldots, SQN[47]$$

The length of SQN is 48 bits. The AuC should include a fresh SQN in each AUTN. Verification of the freshness of the SQN by the USIM constitutes entity authentication of the network to the user.

Parameter AMF is the authentication management field:

$$AMF[0], AMF[1], \ldots, AMF[15]$$

The length of AMF is 16 bits. Its use is not standardized. Example uses of the AMF are provided in Annex F of TS 33.102 [9] (see also Section 2.1.1.1).

Parameter MAC-A and its equivalent XMAC-A are message authentication codes used for authentication of the network to the user:

$$\text{MAC-A}[0], \text{MAC-A}[1], \ldots, \text{MAC-A}[63]$$

The lengths of MAC-A and XMAC-A are both 64 bits. MAC-A is used to authenticate the data integrity and the data origin of RAND, SQN and AMF. The verification of MAC-A by the USIM constitutes entity authentication of the network to the user.

Parameter MAC-S and its equivalent XMAC-S are message authentication codes used to provide data origin authentication for synchronization failure information sent by the USIM to the AuC:

$$\text{MAC-S}[0], \text{MAC-S}[1], \ldots, \text{MAC-S}[63]$$

The lengths of MAC-S and XMAC-S are both 64 bits. MAC-S authenticates the data integrity and data origin of RAND, SQN and AMF. MAC-S is generated by the USIM and verified by the AuC.

Parameter RES and its equivalent XRES are user responses:

$$\text{RES}[0], \text{RES}[1], \ldots, \text{RES}[n-1]$$

The length n of RES and XRES is at most 128 bits and at least 32 bits, and must be a multiple of 8 bits. RES and XRES constitute entity authentication of the user to the network.

Parameter CK is the cipher key:

$$\text{CK}[0], \text{CK}[1], \ldots, \text{CK}[127]$$

The length of CK is 128 bits. In case the effective key length needs to be made smaller than 128 bits, the most significant bits of CK must carry the effective key information, whereas the remaining, least significant bits are set to 0.

Parameter IK is the integrity key:

$$\text{IK}[0], \text{IK}[1], \ldots, \text{IK}[127]$$

The length of IK is 128 bits. In case the effective key length needs to be made smaller than 128 bits, the most significant bits of IK must carry the effective key information, whereas the remaining, least significant bits are set to 0.

Parameter AK is the anonymity key:

$$\text{AK}[0], \text{AK}[1], \ldots, \text{AK}[47]$$

The length of AK is 48 bits (i.e., it equals the length of SQN).

8.2.3 General requirements

The cryptographic functions for AKAs are allocated at the AuC and in the USIM. This means that functions are proprietary to the HE and there is no need for formal standardization of these algorithms. However, the 3G Security Group agreed to develop an example set of functions that could be offered to operators who chose not to develop their own solutions.

The random challenge-generating function f0 is not included in the example set of functions that was to be provided by the authentication function Task Force. Implementation of this function is left to be determined by the operator.

Even if 3GPP AKA algorithms need not be standardized, they are subject to similar requirements as confidentiality and integrity algorithms with respect to resilience and availability.

8.2.3.1 Resilience

Functions need be designed with a view to their continued use for a period of at least 20 years. Successful attacks with a workload significantly less than an exhaustive key search for the subscriber authentication key K must be impossible. Designers of these functions must design sufficiently strong algorithms to reflect the above qualitative requirements.

8.2.3.2 Worldwide availability and use

Legal restrictions on the use or export of equipment containing cryptographic functions may prevent the use of such equipment in certain countries.

It is the intention of the 3GPP that those UE and USIM modules that embody such algorithms should be free from restrictions on export or use, in order to allow the free circulation of 3G terminals. However, network equipment, including the RNC and AuC, may be expected to come under more stringent restrictions. Those RNCs and AuCs that embody such algorithms should be exportable under the conditions of the Wassenaar Arrangement [134].

8.2.4 Additional requirements from SA3

SA3 gave to SAGE the following additional requirements:

- Controlled personalization of the algorithm must be possible based on an operator-variant algorithm configuration field of at least 128 bits.
- The algorithm should be designed around a replaceable kernel function to provide an additional degree of variety.
- If an algorithm is to be designed around a kernel function, then it is required that one specific kernel function be provided.
- If an algorithm is to be designed around a kernel function, then it is desirable that a list of suitable alternative kernel functions be provided.
- If an algorithm is to be designed around a kernel function, then it is desirable that standard/publicly available algorithms be used to implement the kernel function. However, the type or types of kernel function that could be supported is left to SAGE.
- The algorithm must lend itself to implementations that are resistant to Simple Power Analysis (SPA), Differential Power Analysis (DPA) and other "side-channel" attacks as appropriate when implemented on a USIM. It is acknowledged that SAGE may need to consult with smart card experts in order to be able to address this requirement.

8.3 Design Process

8.3.1 Work plan

The algorithm Task Force set up to design an example set of 3GPP AKA functions was formalized in July 2000. On the technical level SAGE decided that the example set of 3GPP AKA functions should be built around a 128-bit block cipher. This was considered an appropriate choice considering the specified parameter lengths and the current standard level of security. Based on this kernel, different modes of operation should be specified, which should include an Operator-variant Parameter (OP).

The algorithms were designed using a phased iterative and interactive approach. A detailed report of the design process is given in [27], which is the main source of the information given in this section and in Section 8.8. In the general report [23] the design process is summarized as follows.

Phase 1 The starting points for algorithm and design criteria were agreed. The design team then produced a first design proposal for a framework algorithm set including different options for the inclusion of OP. A block cipher was proposed as the kernel with Rijndael as the prime candidate.

Phase 2 Results of the first evaluations were discussed and choice of block cipher as kernel versus keyed hash was evaluated. Use of OP as protection against DPA was discussed. Counter mode was adopted instead of OFB. Based on these results, the design team revised the design to a second design proposal for the algorithm set. 3GPP SA3 agreed to limit RES to 64 bits in the example and to change input to the synchronization token.

Phase 3 The results of the second evaluation were discussed and the use of Rijndael and the design for the algorithm set was confirmed, except for some small details. Considering the extensive analyses that had been carried out Rijndael it was decided that the normal range of statistical testing need not be performed. Evaluation of complexity of implementation started.

Phase 4 Details were fixed after mathematical evaluation of the modes and the method for deriving OP_C (see Section 3.5.4.2) from OP was changed. Resistance to side-channel attacks was decided to be left to implementers, who were referred to methods that recommended Rijndael as the candidate to AES. The specification documents were drafted and two parties independently carried out specification testing to check the correctness and completeness of the specification and the accompanying *C*-code.

Phase 5 After a final review the specifications were confirmed. A summary report [27] of the evaluation undertaken by the Task Force was produced and agreed, as well as a general report [23].

The major design goal for the Task Force was to design a framework for AKA functions that was secure and flexible. This goal was achieved by developing of a well-analysed construction using a 128-bit encryption algorithm as a kernel function and including an additional configuration field parameter selected by the operator. The example design recommends the use of the AES algorithm Rijndael as the kernel function, but an operator could change this to any block cipher that meets the interface parameters (the list of candidates for the AES standard includes a large set of suitable algorithms).

The defined set of algorithms is commonly called the "MILENAGE" algorithms (it is not clear what it stands for). An Alta Vista search of the word "milenage" over the Internet in late 2002 gave two responses: the first was a Chardonnay wine by Georges Duboeuf from Pays d'Oc, and the second was a 3GPP algorithm. A wine reviewer at `alcoholreviews.com` writes that Milenage Chardonnay 1999 is well balanced, has soft viscosity and comes in a cleverly designed bottle. If "bottle" were changed to "handset", then these are all qualities that the 3GPP MILENAGE 2000 would be proud to share. Hence one may suspect that the name of the 3GPP authentication algorithm is French in origin, which is confirmed by the instructions given in [24, sect. 0], to pronounce it as a French word—something like "mi-le-nahj".

8.3.2 SAGE's contribution to the UMTS security architecture

When studying 3GPP authentication specifications and algorithm requirements, SAGE made the observation that derivation of the authentication key was essentially different between the regular and the resynchronization case. The same key derivation f5 function was to be used for both purposes, but the inputs were different. In the regular case the inputs were *K* and RAND, while in the resynchronization case the input was MAC-S. This would mean that while the order of execution for all other functions in the USIM is free and even parallel execution would be possible, this would not be allowed when the USIM computes the resynchronization token. In this case, the f1 function should always be computed first. Taking this restriction into account might lead to less efficient implementation.

The reason for such a difference was twofold: first, it was desirable to differentiate cryptographically (called "cryptographic separation") between the regular and the resynchronization case and, second, to prevent an attacker from receiving information of the current SQN_{MS} of the user by sending her outdated RANDs and forcing her to resynchronize. The first reason was considered valid, but the second reason may not be relevant. The reason for concealing SQN and SQN_{MS} is to prevent a certain user being tracked. But if user identity has already been revealed and the user is made to respond to an outdated challenge, then secrecy of the counter values is no longer important.

SAGE proposed a change that would ensure cryptographic separation between the regular and the resynchronization operation, but would allow independent execution of the f1 and f5 functions on the USIM. According to the proposal, inputs to the computation of the AKs would be K and RAND in both cases, but in the resynchronization case a different f5* function would be used.

At the meeting in September 2000, SA3 formally approved the changes proposed by SAGE. In the new version of document [1], the requirement for cryptographic separation was subsequently added as follows: f5* is a key-generating function used to compute the AK in resynchronization procedures that has the property that no valuable information can be inferred from the function values of f5* about those of f1, f1*, f2, . . . , f5 and vice versa.

At the same meeting, SA3 also approved the following changes to the task definition:

- It was not required that the SAGE Task Force produce any example on generating RAND parameter values (f0 function). Random generators typically make use of the various physical features of the implementation environment and a standard example would not be able to exploit such implementation-specific features.

- It was sufficient that the example algorithm family designed by the SAGE Task

Force produce only RES parameter values with lengths in the range 32 to 64 bits (recall that originally a RES length up to 128 bits was allowed).

Finally, SA3 also encouraged SAGE to give the example algorithms to external experts for study even though formal review had been organized.

8.3.3 Cryptographic requirements

Based on the 3GPP algorithm requirement specification [10] (see also Section 8.2) and the set of additional functional requirements described in Section 8.2.4, the Task Force derived a set of cryptographic requirements that authentication functions should comply with. These requirements are given in [27] as follows:

1. Without knowledge of secret keys, the functions f1, f1*, f2, f3, f4, f5 and f5* should be practically indistinguishable from independent random functions of their inputs (RAND || SQN || AMF) and RAND (e.g., knowledge of the values of one function on a fairly large number of given inputs should not enable its values to be predicted on other inputs and the outputs from any one function should not be predictable from the values of the other functions—on the same or other inputs).
2. It should be infeasible to determine any part of the secret key K, or the operator-variant configuration field OP, by manipulation of the inputs and examination of the outputs to the algorithm.
3. Events tending to violate criteria 1 and 2 should be regarded as insignificant if they occur with probability approximately 2^{-128} or less (or require approximately 2^{128} operations).
4. Events tending to violate criteria 1 and 2 should be examined if they occur with probability approximately 2^{-64} (or require approximately 2^{64} operations) to ensure that they do not have serious consequences. Serious consequences would include recovery of a secret key, or ability to emulate the algorithm on a large number of future inputs.
5. The design should build upon well-known structures and avoid unnecessary complexity. This will simplify analysis and avoid the need for a formal external evaluation.

The Task Force also observed that protection against side-channel attacks like DPA may result in increased complexity of the implementation. Therefore it agreed to ensure that the given performance requirements (see Section 8.2.2.4) would be met even after implementation of adequate protection mechanisms ([27], sect. 7.2).

8.3.4 *Operator-variant algorithm configuration field*

In response to a request from SA3, the Task Force decided to include the use of the OP field. This configuration field is used to add operator-dependent information to the design even if the choice of the kernel function is the same. Further, the Task Force identified the following roles of OP ([27], sect. 7.3):

1. To make each operator's implementation different.
2. To prevent USIMs for different operators being interchangeable, either through trivial modification of inputs and outputs or by reprogramming a blank USIM.
3. To keep some algorithm details secret.
4. To provide some protection against a poorly chosen kernel function.

8.3.5 *Criteria for the cryptographic kernel*

The kernel function is used by the MILENAGE framework to produce a 128-bit output value from a 128-bit input value. The output of each of the specific modes (one of the functions f1–f5, f1* or f5*) is further derived from the output of the kernel function. These output values are produced under the control of a 128-bit, user-specific key K. It should be noted that K is a long-term secret that must be protected under all circumstances.

8.3.5.1 Implementation and operational considerations

The kernel function was to become the heart of the design. It had to not only be cryptographically strong it had to be fast as well, since it was to be executed several times to produce the complete output of authentication functions. Therefore it would require the major part of the total memory allocated to authentication functions.

The memory requirements for the full algorithm set was defined as 8 kbytes of ROM and 300 bytes of RAM. From this budget the Task Force decided to allocate at most 6 kbytes of ROM and about 200 bytes of RAM to the kernel function. The total execution time allowed for the full algorithm was 500 ms. The number of calls to the kernel function had still not been determined, but it was estimated that the kernel function should produce a 128-bit output value in less than 50 ms of execution time.

8.3.5.2 Functional requirements

Functional security requirements were defined as the generic requirements for a keyed cryptographic function. The purpose of the kernel function is to map an input value (plaintext) P to an output value (ciphertext) C under the control of a key K. The key must be hidden (i.e., it must be (computationally) infeasible to derive K if an arbitrary amount of pairs P and C are known and K is fixed). It must also be infeasible to compute K by choosing several different P values, applying the kernel function and observing the resulting C values. This kind of chosen plaintext attack must be impossible even if the attacker has access to side-channel information (e.g., power consumption or execution timings of an IC card that holds an implementation of the kernel function).

Furthermore, it must be infeasible to compute C given a P, if K is not known but an arbitrary amount of P and C pairs are known and produced using the same K.

There is no need for the kernel function to be invertible. However, since the input and output values have the same size and collisions are avoided, a bijective function would be a good choice. A justification for choosing an invertible function instead of a hash function is given in Section 8.6.1.

8.3.5.3 Types and parameters for the kernel

The interfaces of the kernel function were also defined at an early stage. The current security standards operate typically on 128-bit data blocks and the minimum length of the key is also 128 bits. A further decision was made that the data input and output blocks should be of equal length, but this was not necessary since hash functions or MAC functions, which do not limit the input size, would also have offered the required functionality.

The sizes of the parameters of the kernel function were defined as follows:

- input block length: 128 bits;
- output block length: 128 bits;
- key length: 128 bits.

Both the key and the input/output blocks are unstructured data (at least from the point of view of the kernel function). It was also noted that the AES candidates are good examples of kernels that meet these requirements [27, sect. 7.4.3].

8.4 Description of the Modes

8.4.1 The algorithm framework

The MILENAGE algorithm set makes use of the following components:

- A block cipher encryption function that takes a 128-bit input and a 128-bit key and returns a 128-bit output.
- A 128-bit value OP. This is an operator-variant algorithm configuration field, which the Task Force was asked to include as a simple means to provide separation between the functionality of the algorithms when used by different operators. It is left to each operator to select a value for OP. The algorithm set is designed to be secure whether or not OP is publicly known; however, operators may see some advantage in keeping their value of OP secret (see also Section 8.7.1).

In the specific example algorithm set, a particular block cipher is used. The algorithms have been designed so that this component is removable and can be replaced by any operator who wishes to create his own customized algorithm set. In that sense MILENAGE defines an algorithm framework, and the example algorithm set is one that fits within the framework. The algorithm set is first defined in terms of an unidentified block cipher and then (see Section 8.6.2) a block cipher is selected to give a fully specified algorithm set.

8.4.2 Notation

The 3GPP MILENAGE algorithm specifications make use of the following conventions: all data variables in this specification are presented with the most significant bit (or byte) on the left-hand side and the least significant bit (or byte) on the right-hand side. Where a variable is broken down into a number of substrings (or single bits), the leftmost (most significant) substring is numbered 0, the next most significant is numbered 1 and so on through to the least significant. The following special notation is used:

$=$ the assignment operator;

$\oplus$ the bitwise XOR operation;

$\|$ the concatenation of two operands;

$E_K(X)$ the result of applying a block cipher encryption to the input value X using the key K;

rot(X, r) the result of cyclically rotating the 128-bit value X by r bit positions toward the most significant bit—if $X = X[0], X[1], \ldots, X[127]$ and $Y = \text{rot}(X, r)$, then $Y = X[r], X[r+1], \ldots, X[127], X[0], X[1], \ldots, X[r-1]$;

$X[i]$ the ith bit of the variable X, $X = X[0], X[1], X[2], \ldots$

This notation is almost identical to the notation used in the specifications, except that here the key K is placed as subscript to the encryption function: as $E_K(X)$ instead of $E[X]_K$.

In addition to the symbols defined for the interface parameters of authentication functions (see Section 8.2.2.6) the algorithm specifications make use of the following symbols:

c_1, c_2, c_3, c_4, c_5 — 128-bit constants that are XORed onto intermediate variables;

IN_1 — a 128-bit value constructed from SQN and AMF and used in the computation of the f1 and f1* functions;

OP — a 128-bit operator-variant algorithm configuration field that is a component of the f1, f1*, f2, f3, f4, f5 and f5* functions;

OP_C — a 128-bit value derived from OP and K and used within the computation of the functions;

$\text{OUT}_1, \text{OUT}_2, \text{OUT}_3, \text{OUT}_4, \text{OUT}_5$ — 128-bit computed values, from which the outputs of the f1, f1*, f2, f3, f4, f5 and f5* functions are obtained;

r_1, r_2, r_3, r_4, r_5 — integers in the range 0–127 inclusive that define amounts by which intermediate variables are cyclically rotated;

TEMP — a 128-bit value used within the computation of the functions.

8.4.3 Specification of the modes

A 128-bit value OP_C is derived from OP and K as follows:

$$\text{OP}_\text{C} = \text{OP} \oplus E_\text{K}(\text{OP})$$

An intermediate 128-bit value TEMP is computed as follows:

$$\text{TEMP} = E_{\text{K}}(\text{RAND} \oplus \text{OP}_{\text{C}})$$

A 128-bit value IN_1 is constructed as follows (recall that SQN is 48 bits and AMF is 16 bits, so concatenating them twice results in a 128-bit value):

$$\text{IN}_1 = \text{SQN} \parallel \text{AMF} \parallel \text{SQN} \parallel \text{AMF}$$

Five 128-bit constants c_1, c_2, c_3, c_4, c_5 are defined as follows:

$$c_1 = 000\ldots00000$$
$$c_2 = 000\ldots00001$$
$$c_3 = 000\ldots00010$$
$$c_4 = 000\ldots00100$$
$$c_5 = 000\ldots01000$$

Five integers r_1, r_2, r_3, r_4, r_5 are defined as follows:

$$r_1 = 64$$
$$r_2 = 0$$
$$r_3 = 32$$
$$r_4 = 64$$
$$r_5 = 96$$

Five 128-bit blocks OUT_1, OUT_2, OUT_3, OUT_4, OUT_5 are computed as follows:

$$\text{OUT}_1 = E_{\text{K}}[\text{TEMP} \oplus \text{rot}(\text{IN}_1 \oplus \text{OP}_{\text{C}}, r_1) \oplus c_1] \oplus \text{OP}_{\text{C}}$$
$$\text{OUT}_2 = E_{\text{K}}[\text{rot}(\text{TEMP} \oplus \text{OP}_{\text{C}}, r_2) \oplus c_2] \oplus \text{OP}_{\text{C}}$$
$$\text{OUT}_3 = E_{\text{K}}[\text{rot}(\text{TEMP} \oplus \text{OP}_{\text{C}}, r_3) \oplus c_3] \oplus \text{OP}_{\text{C}}$$
$$\text{OUT}_4 = E_{\text{K}}[\text{rot}(\text{TEMP} \oplus \text{OP}_{\text{C}}, r_4) \oplus c_4] \oplus \text{OP}_{\text{C}}$$
$$\text{OUT}_5 = E_{\text{K}}[\text{rot}(\text{TEMP} \oplus \text{OP}_{\text{C}}, r_5) \oplus c_5] \oplus \text{OP}_{\text{C}}$$

The outputs of the various f1, f1*, f2, f3, f4, f5and f5* functions are then defined as follows:

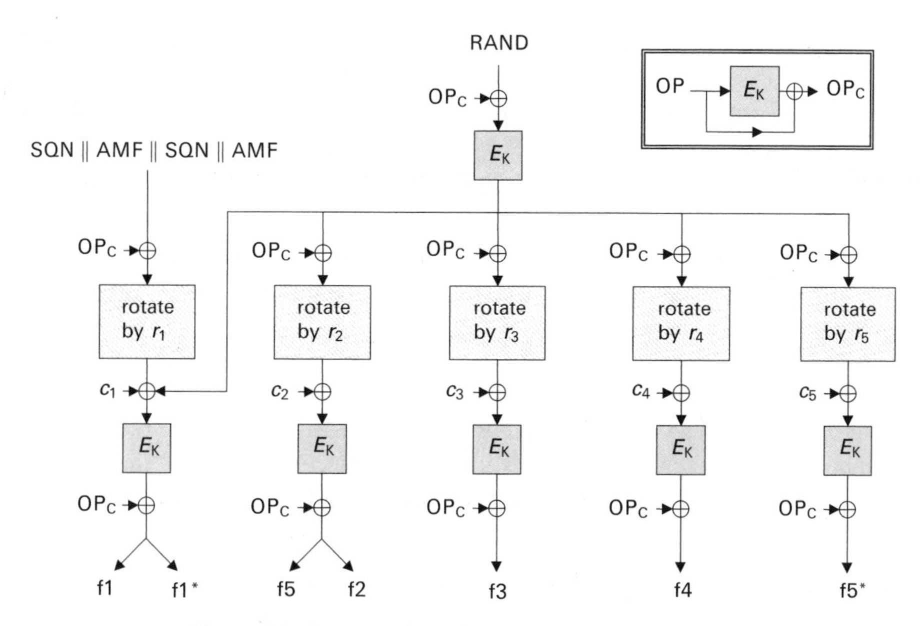

Figure 8.5 Computation of the MILENAGE functions

Output of f1 = MAC-A, where MAC-A[0] ... MAC-A[63] = $OUT_1[0] \ldots OUT_1[63]$
Output of f1* = MAC-S, where MAC-S[0] ... MAC-S[63] = $OUT_1[64] \ldots OUT_1[127]$
Output of f2 = RES, where RES[0] ... RES[63] = $OUT_2[64] \ldots OUT_2[127]$
Output of f3 = CK, where CK[0] ... CK[127] = $OUT_3[0] \ldots OUT_3[127]$
Output of f4 = IK, where IK[0] ... IK[127] = $OUT_4[0] \ldots OUT_4[127]$
Output of f5 = AK, where AK[0] ... AK[47] = $OUT_2[0] \ldots OUT_2[47]$
Output of f5* = AK, where AK[0] ... AK[47] = $OUT_5[0] \ldots OUT_5[47]$

(The repeated reference to AK is not a mistake: AK can be computed as output of either f5 or f5* as these two functions are never computed simultaneously, see Section 8.2.1.)

Computation of the modes is depicted in Figure 8.5. Note that in this figure given by the specification [24] the output of each function is denoted by the function symbol.

8.5 The MILENAGE Architecture

The report [27] also gives the rationale behind the design decisions: in particular, of the use of OP, the selection of various constants and the specific block cipher mode of operation designed for this purpose.

8.5.1 *Use of OP*

The Task Force discussed whether OP should be used directly in the algorithms or whether a derived value should be involved. The Task Force decided to derive a subscriber-dependent value OP_C from OP and the secret key K in a noninvertible way: $OP_C = OP \oplus E_K(OP)$. This construction is noninvertible in both OP and K even if one of them is known. In this case there is no need for storage of OP in each USIM. This means that even if the USIM is compromised, the value of OP could still be kept secret.

The value OP_C is XORed to the input and output of the kernel functions, thus providing additional protection against attacks.

The Task Force recommends that OP_C be computed "off" the USIM as part of the prepersonalization process. This will simplify the algorithms in the card and avoid the storage of OP on the card. It is recommended that OP be kept secret, but MILENAGE is designed to be secure even if the value of OP is known to the cryptanalyst.

An operator could also select different values of OP for different subscribers or subscriber groups.

8.5.2 *Rotation and offset constants*

Rotation constants r_1–r_5 and addition constants c_1–c_5 are selected to ensure separation between all the cryptographic functions involved. The analysis by the Task Force shows that the selected values will protect against collisions in the input (and thus the output) of the final E_K computations (see Section 8.8). If an operator decides to implement other values for these constants, it is strongly advised that the requirements given in the specification are taken into account (see Section 8.7.3).

8.5.3 *Protection against side-channel attacks*

Protection against side-channel attacks is achieved by selection of a kernel that allows for protected implementation within the given time constraints. The surrounding architecture does not provide such protection.

8.5.4 *The number of kernel operations*

For each of the seven functions, the input value RAND passes through two complete rounds of the kernel function before the output values are produced. The encryption of OP in the prepersonalization procedure provides an extra level of security. The other inputs to f1/f1* are obfuscated by XORing them with a random value $E_K(OP_C \oplus RAND)$ and an unknown constant OP_C before they enter the kernel function.

The analysis by the Task Force showed that there are certain forgery attacks against the proposed architecture that involve 2^{64} computations (see Section 8.8). These attacks are not considered feasible within the operational context of 3GPP and would not justify the computational overhead of adding another operation of the kernel function.

8.5.5 Modes of operation

The f1 and f1* constructions are essentially equivalent to the standard CBC MAC mode applied to the input blocks RAND and SQN || AMF || SQN || AMF. The soundness of this construction is theoretically justified by the results in [38].

The f2, f3, f4 and f5 functions are defined as a kind of double encryption of the random challenge with a counter mode construction caused by rotations and constant additions before the second encryption (see Figure 8.6 on p. 232). This type of stateless counter mode is sometimes also called XOR-mode [39]. This is a common mode of operation that has previously been proven secure under certain types of attacks in [39]. Therefore, the soundness of this construction can be assumed to be a direct consequence of the use of a robust kernel function. The Task Force also carried out a rather extensive analysis of the actual construction (the results were reported in [27], see also Section 3.5.8). Recently, a new security proof of the MILENAGE mode of operation was given in [53] within an attack model that is more realistic than the one used in [39].

8.6 Kernel Algorithm

8.6.1 Block ciphers versus hash functions

The functionality provided by the MILENAGE algorithm set was to be twofold: the functionality of MACs and cryptographic check functions was required for f1 and f1*, while other functions were required to be key generation functions. Both functionalities could be provided by building a suitable algorithm framework around a single primitive, which could be either a block cipher or a hash function. The Task Force considered both approaches and the decision was made in favour of a block cipher algorithm. The reasons for the decision are given in [27] as follows.

It was decided to design MILENAGE around a cryptographic kernel function with strong one-wayness properties. To be realistic such an approach requires that well-scrutinized examples of suitable kernel functions are publicly available and preferably on a royalty-free basis. Among the cryptographic functions that would offer the required one-wayness properties two different types can be identified: block ciphers and hash functions. The pros and cons offered by these two alternatives were weighed up against the specific requirements of use and the implementation environment. The decision to select a 128-bit block cipher as a cryptographic kernel was justified as follows:

1. Efficiency of smart card implementation. It is required that the kernel function be efficiently implemented on smart cards with 8-bit processors. The known and commonly used hash functions are all optimized for a larger word size, typically 32 bits.

2. Security of smart card implementation. DPA and other side-channel attacks against smart cards are better understood and analysed in the open literature for certain block ciphers than for any hash functions. Also, protection measures are better developed for certain block cipher structures.

3. Fixed input length. The inputs to the kernel function are parameters of fixed length less than or equal to 128 bits. New block ciphers of a 128-bit block size are suitable for handling such inputs.

4. Secret key input. Block ciphers are designed to take a secret key input. For hash functions such functionality must be constructed artificially. Special keyed modes of operation have been designed for hash functions in the Internet and ISO standards. In ISO 9797 Part 2 [62] three MAC algorithms for dedicated hash functions have been specified. Two of them take at least two applications of the round function of the hash function, which adds extra complexity. One of them, MAC Algorithm 3 is specially designed to take a short maximum 256-bit input and only one application of the round function of the hash function. Of these three MAC algorithms only the Internet HMAC standard is freely available.

5. Availability of block ciphers. There have been many block ciphers around for many years and knowledge about their designs and implementations are well understood and widely known. Even though published 128-bit block ciphers using a 128-bit key have not been around for that many years, the AES process has provided a suite of good candidates for the kernel. On the other hand, there are only a handful of candidate hash functions that are considered secure today.

8.6.2 *The kernel of MILENAGE*

The Task Force agreed to propose the block cipher Rijndael for use as the kernel in the MILENAGE constructions. There were several arguments to support this choice, as quoted from [27]:

- it is a strong encryption algorithm (at the time, it was one of the five AES finalists);
- it is effective and fast on several platforms;
- it is highly suitable for smart card implementation;
- it is freely available without any kind of IPR limitations;
- it can be protected against side-channel attacks;
- it has the required input/output interface;
- it has been published and studied for some time and was based on the design of a previous algorithm called "SQUARE".

In October 2000 Rijndael was chosen as the winner of the AES contest and this should be seen as a strong qualifier for its suitability in the 3GPP environment. In its report [27] the Task Force referred to the AES report [91] where a detailed description of the evaluation and relative merits of Rijndael and the other AES finalists was given and quoted the following text from the conclusion of the AES report by NIST:

> Rijndael appears to be consistently a very good performer in both hardware and software across a wide range of computing environments regardless of its use in feedback or non-feedback modes. Its key setup time is excellent, and its key agility is good. Rijndael's very low memory requirements make it well suited for restricted-space environments, in which it also demonstrates excellent performance. Rijndael's operations are among the easiest to defend against power and timing attacks. Additionally, it appears that some defence can be provided against such attacks without significantly impacting Rijndael's performance.

Later, in November 2001, the 128-bit block version of Rijndael became the AES standard algorithm. The complete specification of the AES algorithm is given in the FIPS (Federal Information Processing Standard) 197 specification [90].

When implementing the AES algorithm for MILENAGE it should be noted that the inputs and outputs of the AES algorithm are defined as strings of bytes. In this way the 128-bit string $X = X[0], X[1], \dots, X[127]$ is treated as a string of bytes by taking $X[0], X[1], \ldots, X[7]$ as the first byte, $X[8], X[9], \ldots, X[15]$ as the second byte and so on. The key and output string are converted in the same way.

For 3GPP applications, the AES is used only in encryption mode and has both block and key length set to 128 bits.

8.7 Customization Options

The specification of 3GPP MILENAGE authentication functions offers operators two main options for customization of the algorithm. These are the freely selectable OP and the removable kernel function. The designers of the MILENAGE set open up a further customization option by allowing the operators to select their own constant and rotation values under certain well-defined conditions (see Section 8.7.3).

8.7.1 Operator variant parameter

The MILENAGE architecture (see Sections 8.4.1 and 8.5.1) allows operators to define their own value for OP that will then be used for their subscribers, leaving operators to decide how they manage OP. Nevertheless, the Task Force report offers the following suggestions:

- the value of OP used for new batches of USIM modules could be changed occasionally;
- a different value could be given to each USIM supplier (the OP could even be given a different value for every subscriber, though the Task Force did not see any advantages in doing so).

Further, the Task Force provides guidelines to show how OP could be implemented. Recall that OP_C is computed from OP and K and that it is only OP_C, not OP, that is ever used in subsequent computations. This gives two alternative options for implementation of the algorithms on the USIM:

1. OP_C computed "off" the USIM—here OP_C is computed as part of the USIM prepersonalization process and is stored on the USIM (OP itself is not stored on the USIM);
2. OP_C computed "on" the USIM—here OP is stored on the USIM (it may be considered as a hardcoded part of the algorithm if preferred) and OP_C is recomputed each time the algorithms are called.

The recommendation of the Task Force is that OP_C is computed off the USIM if possible, since this gives the following benefits (see [24], sect. 5.1]:

- the complexity of the algorithms run on the USIM is reduced;
- it is more likely that OP can be kept secret.

The reasons for the second point are the following. If OP is stored on the USIM, it only takes one USIM to be reverse-engineered for OP to be discovered and published. However, it would be difficult for someone who has discovered even a large number of OP_C and K pairs to deduce OP. This means that the OP_C associated with any other value of K will be unknown, which may make it harder to mount some kinds of cryptanalytic and forgery attacks. The algorithms are designed to be secure whether or not OP is known to the attacker, but a secret OP is one more hurdle in the attacker's path.

8.7.2 Kernel algorithm

The second major customization option offered by the MILENAGE framework is to select a block cipher other than the AES. It is vitally important that whatever block cipher is chosen it is one that has been extensively analysed and is still believed to be secure. The security of authentication and key generation functions is crucially dependent on the strength of the block cipher.

The selection of kernel function is not limited to block ciphers. Since the kernel function need not be invertible, any keyed function (with 128-bit input, key and output) is possible (e.g., a MAC algorithm). The kernel function must satisfy the following cryptographic requirement, as defined by the Task Force:

- Let the key be fixed. Armed with a large number of pairs of chosen input and resulting output, but without initial knowledge of the key, it must be infeasible to determine the key or to predict the output for any other chosen input with probability significantly greater than 2^{-128}.

The Task Force also reminds operators that when selecting a replacement kernel function its suitability and achieved level of protection against side-channel attacks (such as DPA) must be borne in mind.

8.7.3 Rotation and offset parameters

The specification also offers suggestions to operators who wish to customize the algorithms still further. A simple approach is to select different values for the constants c_1–c_5 and r_1–r_5. However, much care must be taken. First, the pairs (c_i, r_i) must all be different (i.e., it must not be allowed that both $c_i = c_j$ and $r_i = r_j$, for any different i and j). For instance, it must not be the case that both $c_2 = c_4$ and $r_2 = r_4$. Second, it is recommended in the specification [24] that the following restrictions are applied:

- c_1 has even parity (i.e., the number of 1's in its binary representation is even);
- c_2–c_5 all have odd parity.

These conditions were derived as a result of the security analysis the Task Force carried out on the MILENAGE framework. However, whatever customization option the operator is planning for, detailed analysis of the consequences of the changes must be carefully carried out from the point of view of security and performance (e.g., if AES is used, the operator might want to use rotation constants that are not multiples of eight, with the intention of improving security by breaking the byte-oriented structure of AES. However, before such a change is implemented one must carefully analyse the consequences this might have on the performance of the algorithm).

8.7.4 Length of RES

The MILENAGE algorithm is designed to support RES lengths from 32 to 64 bits. Its f2 function produces 64-bit output, which can be used to derive the RES value. The method of derivation is not specified by MILENAGE but left to operators, who use different methods. Some of the most typical methods are discussed in Section 8.8.1.

8.8 Conversion to and Compatibility with A3/A8

The GSM network uses two related algorithms, A3 and A8, to authenticate the subscriber (A3) and generate an encryption key (A8). Like the UMTS f1–f5 algorithms, the A3/A8 algorithms are also operator-specific.

When a UMTS terminal visits a network that only supports GSM authentication and encryption, the terminal must authenticate itself using the A3/A8 interface. Section 8.8.1 describes the conversion rules that are used to convert the A3/A8 inputs to f1–f5 inputs and the outputs of f1–f5 to A3/A8 outputs.

These conversion rules can also be used with MILENAGE algorithms to implement GSM SIM cards, instead of relying on proprietary algorithms or the older

GSM example algorithms COMP128-1 and COMP128-2. This case is discussed in Section 8.8.2.

8.8.1 *Conversion rules*

Specification [9] defines rules by means of which a UMTS authentication quintet can be converted to a GSM authentication triplet. Such a conversion becomes necessary when the visiting network only supports GSM (Release 1998) authentication and encryption.

The A3 authentication algorithm takes a 128-bit challenge RAND as input and produces a 32-bit response SRES, using a 128-bit, long-term key Ki. The A8 key generation algorithm takes the 128-bit challenge RAND as input and produces a 64-bit ciphering key Kc, using the 128-bit, long-term key Ki.

Three conversion rules, denoted by c1, c2, and c3, are needed to derive the GSM triplet values denoted by $\text{RAND}_{[\text{GSM}]}$, $\text{SRES}_{[\text{GSM}]}$ and $\text{Kc}_{[\text{GSM}]}$ for GSM:

$$\text{c1}: \text{RAND}_{[\text{GSM}]} = \text{RAND}$$

$$\text{c2}: \quad \text{SRES}_{[\text{GSM}]} = (\text{XRES}^*[0]\ldots\text{XRES}^*[31]) \oplus (\text{XRES}^*[32]\ldots\text{XRES}^*[63]) \oplus (\text{XRES}^*[64]\ldots\text{XRES}^*[95]) \oplus (\text{XRES}^*[96]\ldots\text{XRES}^*[127])$$

$$\text{c3}: \quad \text{Kc}_{[\text{GSM}]} = (\text{CK}[0]\ldots\text{CK}[63]) \oplus (\text{CK}[64]\ldots\text{CK}[127]) \oplus (\text{IK}[0]\ldots\text{IK}[63]) \oplus (\text{IK}[64]\ldots\text{IK}[127])$$

where XRES* is 16 octets long and derived from XRES as follows: if XRES is 16 octets long, XRES* = XRES; if XRES is shorter than 16 octets, it is padded with 0 bits.

When a GSM subscriber is attached to a UTRAN, the authentication is based on the SIM card. Then the 128-bit UMTS cipher and integrity keys $\text{CK}_{[\text{UMTS}]}$ and $\text{IK}_{[\text{UMTS}]}$ are derived from the 64-bit GSM cipher Kc key using the following conversion functions:

$$\text{c4}: \text{CK}_{[\text{UMTS}]} = \text{Kc}[0]\ldots\text{Kc}[63] \,\|\, \text{Kc}[0]\ldots\text{Kc}[63]$$

$$\text{c5}: \quad \text{IK}_{[\text{UMTS}]} = (\text{Kc}[0]\ldots\text{Kc}[31]) \oplus (\text{Kc}[32]\ldots\text{Kc}[63]) \,\|\, \text{Kc}[0]\ldots\text{Kc}[63] \,\|\, (\text{Kc}[0]\ldots\text{Kc}[31]) \oplus (\text{Kc}[32]\ldots\text{Kc}[63])$$

The conversion rules c3, c4 and c5 have been chosen to have the following property: given a GSM cipher key Kc, if we first apply conversion rules c4 and c5 to derive CK and IK for UMTS and afterward apply conversion rule c3, then the original Kc is returned as $\text{Kc}_{[\text{GSM}]}$.

8.8.2 GSM–MILENAGE

8.8.2.1 Compatibility with conversion rules

The 3GPP algorithm Task Force was also asked to derive GSM A3 and A8 algorithms based on MILENAGE. The straightforward way of carrying out this task is just to define the conversion interface on MILENAGE. However, this is not completely trivial due to the fact that the MILENAGE specification leaves it to the operator to define how the 64-bit output of the MILENAGE f2 function is truncated if the operator's system uses shorter RES and XRES values.

The following three options are arguably the most typical ways of using RES and XRES values by different operators:

1. RES(option 1) = RES[0] ... [63];
2. RES(option 2) = RES[0] ... RES[31];
3. RES(option 3) = (RES[0] ... RES[31]) $\oplus$ (RES[32] ... RES[63])

where RES denotes the 64-bit output of the MILENAGE f2 function (see Section 8.4.3). Clearly, if the conversion rule c2 is applied to RES options 1 or 3, the result would be the same, but a different result is obtained when the conversion rule is applied to RES option 2. It is for this reason that the GSM–MILENAGE specification proposes two alternatives to support these three UMTS RES derivation methods. If an operator has selected a UMTS RES derivation method that is different from these three options, then to achieve compatibility with the conversion rules a proprietary RES derivation function may have to be defined.

8.8.2.2 The GSM–MILENAGE A3 and A8 algorithms

The GSM–MILENAGE A3 algorithm is defined using the MILENAGE f2 functions, while definition of the GSM–MILENAGE A8 algorithm makes use of the MILENAGE f3 and f4 functions (the specification is given in [30]).

For the purposes of defining the GSM–MILENAGE specification all input and output names of the UMTS f2, f3 and f4 algorithms (see Section 8.2.2.6) are modified by adding the prefix "MIL3G" to distinguish them clearly from the inputs and outputs of A3 and A8. In this way, the inputs K and RAND to the f2, f3 and f4 functions are denoted as:

$$\text{MIL3G-K} = \text{MIL3G-K}[0]\ldots\text{MIL3G-K}[63]$$

and

$$\text{MIL3G-RAND} = \text{MIL3G-RAND}[0]\ldots\text{MIL3G-RAND}[127]$$

The output RES of f2 is denoted as:

$$\text{MIL3G-RES} = \text{MIL3G-RES}[0] \ldots \text{MIL3G-RES}[63]$$

The output CK of f3 is denoted as:

$$\text{MIL3G-CK} = \text{MIL3G-CK}[0] \ldots \text{MIL3G-CK}[127]$$

The output IK of f4 is denoted as:

$$\text{MIL3G-IK} = \text{MIL3G-IK}[0] \ldots \text{MIL3G-IK}[127]$$

Then the GSM–MILENAGE functions are defined as follows:

- Set MIL3G-K = Ki, where Ki is the GSM subscriber key.
- Set MIL3G-RAND = RAND, where RAND is the GSM RAND.
- Compute MIL3G-RES, MIL3G-CK and MIL3G-IK from MIL3G-K and MIL3G-RAND, using the MILENAGE f2, f3, and f4 functions, respectively.
- Set

$$\begin{aligned}\text{Kc} = (\text{Kc}[0] \ldots \text{Kc}[63]) &= (\text{MIL3G-CK}[0] \ldots \text{MIL3G-CK}[63]) \\ &\oplus (\text{MIL3G-CK}[64] \ldots \text{MIL3G-CK}[127]) \\ &\oplus (\text{MIL3G-IK}[0] \ldots \text{MIL3G-IK}[63]) \\ &\oplus (\text{MIL3G-IK}[64] \ldots \text{MIL3G-IK}[127])\end{aligned}$$

- Derive SRES from MIL3G-RES using an operator-selected *SRES derivation function*, which must be precisely specified for the GSM–MILENAGE A3 algorithm to be fully defined. The two main recommended options are as follows:

  1. Recommended SRES derivation function #1:

$$\begin{aligned}(\text{SRES}[0] \ldots \text{SRES}[31]) &= (\text{MIL3G-RES}[0] \ldots \text{MIL3G-RES}[31]) \\ &\oplus (\text{MIL3G-RES}[32] \ldots \text{MIL3G-RES}[63])\end{aligned}$$

  2. Recommended SRES derivation function #2:

$$(\text{SRES}[0] \ldots \text{SRES}[31]) = (\text{MIL3G-RES}[0] \ldots \text{MIL3G-RES}[31])$$

Consider the three methods identified in Section 8.8.2.1 in which the UMTS RES value can be derived from the 64-bit output of MILENAGE. If method 1 or 3 is used in UMTS, the recommended SRES derivation function #1 gives a result compatible with conversion rule c2. If method 2 is used in UMTS (i.e., the 32-bit UMTS RES is obtained from the 64-bit output of MILENAGE by just truncating it), then SRES derivation function #2 is recommended, as it gives the same result as the conversion rule c2.

8.8.2.3 An alternative algorithm

These GSM–MILENAGE A3 and A8 algorithms are obtained by applying standard UMTS-to-GSM conversion functions to the outputs of the UMTS–MILENAGE algorithms.

If there is no desire to retain this compatibility with UMTS–MILENAGE used in its GSM mode, a simpler and more efficient algorithm can be obtained by using any robust block cipher algorithm of 128-bit block size and 128-bit key. Here we would set TEMP = E_{Ki}(RAND), where E is the encryption transformation of the block cipher (i.e., TEMP = the result of encrypting RAND using the key Ki) and then set SRES = TEMP[0] ... TEMP[31] and Kc = TEMP[64] ... TEMP[127].

This alternative does *not* form any part of the GSM-MILENAGE algorithms; it is included in the new A3 and A8 specification prepared by the Task Force for information purposes only.

8.9 Security Analysis of MILENAGE

8.9.1 Assumptions and security claims

The starting point of the security evaluation carried out by the Task Force was that the kernel function needed to be a robust block cipher of 128-bit block size and 128-bit key length. With this in place, it remained to show that the MILENAGE authentication functions satisfy predefined security requirements under the assumption that a strong block cipher encryption function had been used as the kernel function in the MILENAGE framework.

A strong block cipher (as used in the analysis) was defined as:

> A 128-bit block cipher E_K is said to be a *strong block cipher*, if there is no efficient test that allows us to distinguish permutation E_K for a randomly drawn key K from a randomly drawn permutation of the set $\{0, 1\}^{128}$, with substantially less effort than 2^{128} data (encryption or decryption results) and significantly less than 2^{128} E_K operations.

The AES block cipher underwent extensive analysis during the AES process [91].

Hence the Task Force decided not to duplicate this work or carry out its own cryptanalytical research on the Rijndael cipher. In particular, it decided not to carry out any statistical tests on the MILENAGE functions, since any such tests would have been performed on MILENAGE, would only yield results about the underlying kernel function (see [27, sect. 10.4]).

Instead, the focus of the cryptanalysis was to assess the security of the MILENAGE construction for deriving different f1–f5* function modes from a strong block cipher. To do this, mathematical evaluation needs to consider, first, the strength of each function mode f1–f5* individually and, second, the independence of the different f1–f5* function modes. These security requirements were formalized as follows:

1. There must be no attack that takes substantially less than 2^{128} computations of E_K to successfully recover any information on the value of the key K or to forge outputs of the algorithm for a substantial set of arbitrary RAND inputs, based on knowledge of algorithm outputs that correspond to any chosen RAND, SQN, AMF inputs, even if the OP, c_i and r_i values are known.

2. There must be no other attack that enables the seven function generators $K \mapsto$ f1*, f2, f3, f4, f5, f5* to be distinguished from independent random functions of the 192-bit input RAND || SQN || AMF (f1 and f1* modes) or the 128-bit input RAND (other modes) using substantially less than 2^{64} queries, even if the OP, c_i and r_i values are known. In particular, given any combination of f1–f5* function significantly smaller than 2^{64} outputs related to any chosen inputs, it must be computationally infeasible to predict any additional output for any of the fi or fi* functions—even if outputs corresponding to the same RAND value are known for the other modes.

The mathematical evaluation carried out by the Task Force came up with three different types of results. First, formal arguments were presented to validate some aspects of the algorithm construction. Second, more informal security arguments were presented on other aspects of the construction not covered by the formal arguments and, third, various types of forgery and distinguishing attacks having data or time complexity close to the bound of 2^{64} were studied.

8.9.2 Operational context

The MILENAGE construction is a special purpose set of algorithms designed for use in a well-defined operational environment. Therefore, its security was also analysed only within its operational context, which means that some attacks may be easy to launch, while others are impossible or prevented by some other means. For example, it was considered that attacks based on known relations between subscriber keys are impossible in this operational context or at worst would not produce any useful information (for the attacker).

On the other hand, all attacks exploiting direct access to the USIM module must be very carefully analysed. In a direct attack on USIM we must assume that:

- an attacker has full control over the values of the RAND, SQN and AMF inputs;
- the output of f1 (i.e., MAC-A) is checked within the USIM and is not directly available to an attacker, but all other outputs are available to the attacker;
- the input and output bandwidth of the USIM is limited, as is its processing power, and as a result the practical rate at which input and output pairs can be collected is severely limited.

The estimate given in the report (i.e., 10 pairs or less per second) is certainly on the safe side taking into account that the total time allowed for the USIM to process one pair is 500 ms.

8.9.3 *The soundness of the f2–f5* construction*

The most distinctive features of the modes for the f2–f5* functions are captured in the simplified model depicted in Figure 8.6. Parameter t is fixed and denotes the

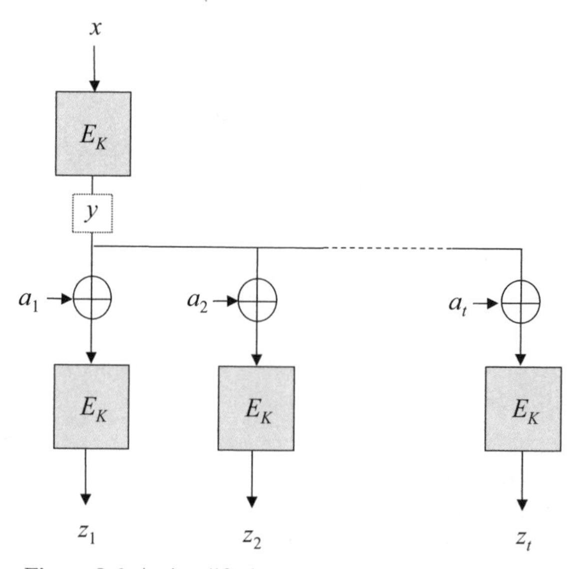

Figure 8.6 A simplified counter mode construction

number of distinct output blocks. The values $a_1, a_2, \ldots, a_t$ are assumed to be any t fixed, known, distinct offset constants. The offset a_i is applied to the input y before it is taken as input to the encryption function of the ith block. For the MILENAGE construction of the f2 to f5* functions, the value t is equal to 4.

The aim is to prove there is no way to use any combination of significantly less than 2^{64} output values $z_1, z_2, \ldots, z_t$ to predict any new output value for any of the z_i blocks. More generally, it needs to be shown that if E_K behaves as a random permutation of the $\{0,1\}^n$ set, then the function that maps x to the t-tuple $z_1, z_2, \ldots, z_t$ cannot be distinguished from any random function between $\{0,1\}^n$ and $\{0,1\}^{n \cdot t}$ in any efficient way.

To do this, the occurrences of the E_K function in Figure 8.6 are replaced by a perfectly random permutation (i.e., a uniformly drawn element from the set of permutations of the set $\{0,1\}^n$). The construction then becomes a random function f between $\{0,1\}^n$ and $\{0,1\}^{n \cdot t}$. Now, the idea is to compare this function f with a uniformly drawn random function f^* between $\{0,1\}^n$ and $\{0,1\}^{n \cdot t}$.

The comparison is based on an arbitrary distinguishing algorithm (distinguisher) A of unlimited power. The distinguisher is given a black box that contains a function, which has been generated either as f or f^*. Then the distinguisher is allowed to make a fixed number of queries about the output values of the function for distinct, chosen or adaptively chosen input values. After seeing the results, the distinguisher outputs one bit value: 0 or 1. Let p be the probability that A outputs 1 when the given function is of type f and let p^* be the probability that A outputs 1 when fed with a type of f^*.

It is said that a construction of type f cannot be distinguished from a perfectly random function f^* if, for any distinguisher A, the probabilities p and p^* are about the same. The absolute value $|p - p^*|$ is called the advantage of distinguisher A and is denoted by $\mathrm{Adv}_A(f, f^*)$.

In [27], the Task Force proved a result that confirms that the construction of type f cannot be distinguished from a perfectly random function f^* with essentially less than $2^{n/2}$ queries. The result is formulated in [27] along these lines:

> Let n be any fixed integer and c^* any perfectly random permutation of the set $\{0,1\}^n$. Let $f = \Phi(c^*)$ denote a random function between $\{0,1\}^n$ and $\{0,1\}^{n \cdot t}$ obtained by applying the countermode construction of Figure 8.6 to c^* in place of E_K, and let f^* denote a perfectly random function between $\{0,1\}^n$ and $\{0,1\}^{n \cdot t}$. For any distinguishing algorithm A that uses a fixed number q of queries we have $\mathrm{Adv}_A(f, f^*) \leq 3t^2q^2/2^{n+1}$.

Further, it is concluded in [27] that the result can also be adapted to accommodate rotations and that the involvement of OP_C does not seem to degrade security in any extensive way. Thus, in summary, the design of the f2–f5* functions appears to be sound and to comply with the design criteria.

The above theorem was improved and equipped with a proof by Henri Gilbert in [53]. The new formulation of the theorem also incorporates offset constants and the advantage bound is decreased by a factor of 3 to $t^2q^2/2^{n+1}$.

Previously, a similar "one-block-to-many" mode of operation, like the one depicted in Figure 8.6, was studied by Bellare et al. in [39], where it was called the XOR-mode. However, the XOR mode was defined as an encryption mode of operation and its security was only evaluated for this functionality. In the security model used in [39] the attacker is only allowed to choose plaintext or access ciphertext, but is assumed to have no access to the input of the keystream generator. For the MILENAGE construction, however, it is also essential to prove security when the attacker can choose random challenges and other input parameters.

In the model used in [39], distinguishability is limited to a special type of distinguisher that attempts to determine, given two plaintexts and an encryption, which of the plaintexts was encrypted using the keystream generated by the function. This distinguishability notion is known as "left-or-right distinguishability" (see also Section 6.6.3.2).

The use of different attack models may also explain the essential difference in the security bounds between [39] and [27]. Both bounds are quadratic in the number of queries, but the bound given in [39] is only a linear function of t.

8.9.4 Soundness of the f1–f1 construction and its cryptographic separation from the other modes*

The analysis presented in [27] goes into the details of the actual design. The various values used in MILENAGE are defined in Sections 8.2.2.6 and 8.4.2, but for the purposes of the mathematical analysis presented in this and the following section the following simpler notation is introduced:

$$
\begin{aligned}
x &= \mathrm{RAND} \\
y &= \mathrm{TEMP} = E_{\mathrm{K}}(x \oplus \mathrm{OP_C}) \\
u &= \mathrm{SQN} \,\|\, \mathrm{AMF} \,\|\, \mathrm{SQN} \,\|\, \mathrm{AMF} \\
w_i &= \mathrm{OUT}_i, \quad \text{for } i = 1, 2, 3, 4, 5.
\end{aligned}
$$

All previous notation is kept the same.

8.9.4.1 Soundness of the f1–f1* construction

The f1 and f1* constructions are essentially similar to the standard CBC-MAC construction applied to the message $x \,\|\, u$, with a final truncation of the CBC computation output. Moreover, it is noted that f1 and f1* use distinct output bits, so that cryptographic separation between f1 and f1* appears to be sufficient.

A formal proof of the soundness of the standard CBC-MAC was first established by Bellare et al. [38]. It is argued in [27] that it is possible to transpose the results of [38] to the MILENAGE f1–f1* functions using techniques similar to those applied to the f2–f5* functions and to prove that the f1–f1* construction cannot be distinguished from a random function using substantially less than 2^{64} queries.

On the other hand, it is noted in [27] that there exists a simple, internal collision attack against the standard CBC-MAC that requires about 2^{64} queries, which means that the bound 2^{64} (i.e., the lower bound to the number of queries needed by a distinguisher derived by the result of Bellare et al.) is also an upper bound. This internal collision attack can be transposed to the f1 and f1* functions (see Section 8.9.5.1), but due to the large number of required RAND values this attack is only of theoretical importance.

It is concluded in [27] that the MILENAGE f1–f1* construction appears to be sound, with the only weakness being that impractical attacks with 2^{64} queries were identified.

8.9.4.2 Cryptographic separation between f1–f1* and f2–f5*

Cryptographic separation between the pair f1 and f1* and the remaining functions f2–f5* is based on the fact that the offset constant c_1 has even parity, while all the other constants c_2–c_5 have odd parity. This will be shown next.

According to the definition of output blocks w_i, their functional expressions are as follows:

$$w_1(x,u) = \mathrm{OP_C} \oplus E_\mathrm{K}(y \oplus c_1 \oplus \mathrm{rot}(u \oplus \mathrm{OP_C}, r_1))$$

$$w_i(x) = \mathrm{OP_C} \oplus E_\mathrm{K}(c_i \oplus \mathrm{rot}(y \oplus \mathrm{OP_C}, r_i))$$

Since the outputs of f1 and f1* are derived from w_1 and the outputs of the remaining functions are derived from w_2, w_3, w_4 or w_5, it is important to look at the connection between w_1 and w_i, where $i = 2$, 3, 4 or 5. Any such connection could possibly be exploited by an adversary who is able to predict events of the form $w_1(x, u) = w_i(x')$.

It was noted in [27] that events of the form $w_1(x, u) = w_i(x)$ (i.e., collisions of output blocks that involve the same random challenge value x) have to be looked at with particular care because:

- Despite the fact that y is unknown, the equation $w_1(x, u) = w_i(x)$, which can be rewritten as $y \oplus \mathrm{rot}(y, r_i) = c_1 \oplus c_i \oplus \mathrm{rot}(u, r_1) \oplus \mathrm{rot}(\mathrm{OP_C}, r_1) \oplus \mathrm{rot}(\mathrm{OP_C}, r_i)$, provides some partial information on $\mathrm{OP_C}$, because $y \oplus \mathrm{rot}(y, r_i)$ is at least partially known and even entirely known if $r_i = 0$ (in which case it is equal to 0).

- If an event of the form $w_1(x, u) = w_i(x)$ occurs for a particular value (x, u), then $w_1(x', u) = w_i(x')$ still holds for any x' value such that the corresponding $y' = E_K(x' \oplus \mathrm{OP_C})$ satisfies $y' \oplus \mathrm{rot}(y', r_i) = y \oplus \mathrm{rot}(y, r_i)$. Thus, if $r_i = 0$ and there is a particular value u such that $w_1(x, u) = w_i(x)$ for some value x, then $w_1(x, u) = w_i(x)$ for all x.

The offset constants c_1–c_5 were selected in such a way as to avoid the event $w_1(x, u) = w_i(x)$ and, subsequently, to avoid any weakness that might result from this event. In fact:

- c_1 is a 128-bit word of even parity;
- u is obtained by repeating the 64-bit word SQN || AMF twice, so u has even parity;
- all other offset constants c_2–c_5 have odd parity.

Taking into account the fact that rotations do not affect the parity of any word, it can be seen that the parity of $E_K^{-1}(w_1 \oplus \mathrm{OP_C}) = y \oplus c_1 \oplus \mathrm{rot}(u \oplus \mathrm{OP_C}, r_1)$ is equal to the parity of $y \oplus \mathrm{OP_C}$, whereas the parity of $E_K^{-1}(w_i \oplus \mathrm{OP_C}) = c_i \oplus \mathrm{rot}(y \oplus \mathrm{OP_C}, r_i)$ is the inverse of the parity of $y \oplus \mathrm{OP_C}$, so that the event $w_1(x, u) = w_i(x)$ can never occur.

In short, then, criteria on the c_1–c_5 and r_1–r_5 defined in the MILENAGE specification seem to ensure an effective separation of f1–f1* functions from f2–f5* functions.

8.9.5 *Investigation of forgery or distinguishing attacks with 2^{64} queries*

Finally, the evaluation report [27] considers the most important analytical methods that could be launched on MILENAGE using 2^{64} known or chosen data. First, it must be noted that since the E_K functions are permutations, the output blocks w_i are computed using bijective functions of the input, $x = \mathrm{RAND}$. In particular, the MILENAGE f3 and f4 functions are invertible functions of RAND. If, say, f3 was a perfectly random function, there would be at least two RAND values in a set of about 2^{64} different RAND inputs that would produce the same f3 output CK. Hence f3, and similarly f4, can be distinguished from a random function using about 2^{64} data. This distinguishing method is trivial in the sense that is follows directly from the choice of using a block cipher encryption algorithm as the kernel function. This choice is not known to create any vulnerabilities and may actually be advantageous in that the produced cipher keys of different RAND values are different.

The analytical methods (attacks) identified and reported in [27] can target a single function (namely f1 or f1*) or combinations of several functions and require about 2^{64} queries (i.e., known or chosen data blocks).

8.9.5.1 An internal collision attack against f1 (or f1*)

As already mentioned in Section 8.9.4.1, the well-known internal collisions of CBC-MAC can also be found for the MILENAGE f1 function.

Consider a set of about 2^{64} (x, u) pairs such that all values of x are distinct and all values of u are distinct. In this case the corresponding 64-bit outputs z_1 of function f1 need to be investigated. According to the Birthday Paradox, it may then be possible with large probability to find two pairs (x', u') and (x'', u'') such that the two corresponding 128-bit output words $w_1(x', u')$ and $w_1(x'', u'')$ collide. Since f1 gives out only half of the output w_1 output block, such a collision cannot be verified directly and, so, more analysis is needed.

A collision $w_1(x', u') = w_1(x'', u'')$ occurs exactly when:

$$y' \oplus \mathrm{rot}(u' \oplus \mathrm{OP_C}, r_1) = y'' \oplus \mathrm{rot}(u'' \oplus \mathrm{OP_C}, r_1)$$

The observation was made that if a δ value is now XORed on both sides, the equation remains valid. This means that the inputs u' and u'' to w_1 computation can be varied using δ values in such a way that the following equation holds:

$$w_1(x', u' \oplus \delta) = w_1(x'', u'' \oplus \delta)$$

This property can be used to detect collisions and afterward to actually forge new z_1 values.

In the generated set of 2^{64} $\mathrm{f1}(x, u)$ values, the cryptanalyst can expect to find many collisions, so-called partial collisions, and just one value is expected to originate from a full collision with the whole w_1 block. To detect a full collision, we can test each of the partial collisions $\mathrm{f1}(x', u') = \mathrm{f1}(x'', u'')$ to see whether $\mathrm{f1}(x', u' \oplus \delta) = \mathrm{f1}(x'', u'' \oplus \delta)$ as well. For a full collision this holds for all δ, while a wrong pair is expected to fail the test after trying one or two values of δ. In this manner a full collision is found in less than 2^{64} operations. Once a full collision (i.e., a pair of inputs (x', u') and (x'', u'') such that $w_1(x', u') = w_1(x'', u'')$) has been identified, it is possible to produce the $\mathrm{f1}(x'', u'' \oplus \delta)$ value based on a given $\mathrm{f1}(x', u' \oplus \delta)$.

Clearly the same collision analysis holds for f1*. f1 and f1* functions are essentially similar to the standard CBC-MAC for which such a collision analysis method is known to exist. Since the method requires about 2^{64} f1 (or f1*) outputs corresponding to distinct RAND inputs, it has no practical significance. Therefore, the use of a MAC mode that is essentially equivalent to the standard CBC mode seems appropriate in the 3GPP context of operation. Moreover, as noted in [27], it would be possible to prevent this type of collision analysis method by introducing an additional E_K block in the construction. However, such a modification would have increased the computational overhead significantly.

8.9.5.2 Forgery or distinguishing attacks against combinations of several modes

For some particular values of rotation constants r_1–r_5 and the offset constants c_1–c_5 some attacks were found on combinations of the f1–f5* functions, which required about 2^{64} queries. These are described below.

Attacks against combinations of f2–f5

Two of the one-block outputs w_2 to w_5, corresponding to two equal or distinct random challenges x' and x'', are equal (i.e., $w_i(x') = w_j(x'')$), exactly when:

$$\mathrm{rot}(y', r_i) \oplus \mathrm{rot}(y'', r_j) = c_i \oplus c_j \oplus \mathrm{rot}(\mathrm{OP_C}, r_i) \oplus \mathrm{rot}(\mathrm{OP_C}, r_j)$$

We now identify particular cases in which a simple attack requiring about 2^{64} queries exists.

Case 1 $r_i = r_j$. In this case the following forgery attack holds. Assume that 2^{64} inputs x and corresponding outputs $w_i(x)$ and $w_j(x)$ are given. There is a large probability that two input values x' and x'' exist such that $w_i(x') = w_j(x'')$. It is easy to see, by the above equation and the fact that $r_i = r_j$, that we also have $w_i(x'') = w_j(x')$. In other words, if an adversary finds two x' and x'' inputs such that $w_i(x') = w_j(x'')$ and obtains the value $w_i(x'')$, then the adversary can also obtain $w_j(x')$. Such a phenomenon would be extremely unlikely to happen if w_i and w_j were the outputs of two independent permutations of x.

Case 2 $r_i - r_j = 64 \bmod 128$, and there is a value v such that $c_i \oplus c_j = \mathrm{rot}(v, r_i) \oplus \mathrm{rot}(v, r_j)$. For given r_i and r_j, with $r_i - r_j = 64$, the set of 128-bit blocks of the form $\mathrm{rot}(v, r_i) \oplus \mathrm{rot}(v, r_j)$ constitutes a 64-dimensional vector subspace of $\{0, 1\}^{128}$. For instance, if $r_i = 0$ and $r_j = 64$, then this subspace is formed by 128-bit blocks with two equal 64-bit halves. Hence, if $c_i \oplus c_j = \mathrm{rot}(v, r_i) \oplus \mathrm{rot}(v, r_j)$ for some v, then the same holds for 2^{64} different values of v. In this case the following distinguishing attack holds. Assume that 2^{64} inputs x and the corresponding outputs $w_i(x)$ and $w_j(x)$ are given. There is a large probablity for at least one of these x values that $w_i(x) = w_j(x)$. As a matter of fact, this equality occurs if and only if $\mathrm{rot}(y \oplus \mathrm{OP_C}, r_i) \oplus \mathrm{rot}(y \oplus \mathrm{OP_C}, r_j) = c_i \oplus c_j$ and there are 2^{64} possible $y \oplus \mathrm{OP_C}$ values that satisfy that equation. Such an event would be extremely unlikely to occur were w_i and w_j the outputs of two independent random permutations.

Attacks against combinations of f1–f1 and f2–f5**

Input values x', u and x'' produce a collision between w_1 and w_i, where $i = 2, 3, 4$ or 5, if $w_1(x', u) = w_i(x)$. This happens exactly if:

$$y' \oplus \text{rot}(y'', r_i) = c_1 \oplus c_i \oplus \text{rot}(u, r_1) \oplus \text{rot}(\text{OP}_\text{C}, r_1) \oplus \text{rot}(\text{OP}_\text{C}, r_i)$$

We have already considered such a situation in Section 8.9.4.2, where it was shown that the parities of the offset constants c_i prevent this equation from being satisfied if $x' = x''$. Nevertheless, in some particular cases, there remain simple attacks that require about 2^{64} queries.

Case 3 $r_i = 0$. In this case the following forgery attack holds. Assume that 2^{64} inputs (x, u) and the corresponding outputs $w_1(x, u)$ and $w_i(x)$ are given. There is a large probability that x', u' and x'' exist such that $w_1(x', u) = w_i(x'')$. It then follows from this equation and the fact that $r_i = 0$ that we also have $w_1(x'', u) = w_i(x')$, allowing us to forge $w_i(x')$ based on our knowledge of $w_1(x'', u)$. Had only outputs of f1 or f1* been available, instead of entire w_1 outputs, the above property would still make it possible for us to distinguish the MILENAGE construction from a random function using about 2^{64} queries.

Case 4 $r_1 = 0$ or 64, and there exist two distinct values $i, j \in \{2, 3, 4, 5\}$ such that $c_i \oplus c_j$ consists of two equal 64-bit halves and $r_i = r_j$. In this case the following forgery attack holds. Assume that 2^{64} inputs (x, u) and the corresponding outputs $w_1(x, u)$ and $w_i(x)$ are given. There is a large probability that x', u_i and x'' exist such that $w_1(x', u_i) = w_i(x'')$. Let us now replace the u_i sequence with the sequence $u_j = u_i \oplus c_i \oplus c_j$, which has equal halves and can therefore be constructed as SQN || AMF || SQN || AMF. It then follows that $w_1(x', u_j) = w_j(x'')$ also holds. In other words, a collision between a w_1 and a w_i value allows us to predict a collision between a w_1 value and a w_j value. Had only outputs of f1 or f* been available, instead of entire w_1 outputs, the above property would still lead to a distinguishing attack using about 2^{64} queries.

Conclusion from the forgery and distinguishing attacks

In all these cases with special constant values attacks can be prevented by appropriate selection of constants. On the other hand, all attacks are highly unlikely, since they require outputs for known or chosen random challenges as well as AMF and SQN values. The main purpose of analysing such situations and combinations of constants is to gain some theoretical information about the structure and relation between different function modes.

It should also be noted that the internal collisions on the MILENAGE f1 and f1* functions described in Section 8.9.5.1 are inherent in these types of constructions.

The attacks based on the specific constant values discussed above do not pose any greater threat than that of an internal collisions attack. The selected values of the offset and rotation constants for MILENAGE prevent all other attacks except the one identified in Case 3, which remains valid for $i = 2$ as $r_2 = 0$.

8.9.6 Conclusions

The cryptographic security requirements for the UMTS AKA algorithm were set in Section 8.3.3. Of the five given requirements the first two define the basic cryptographic attacks that the algorithm should withstand. Requirements 3 and 4 define the security level and lower bounds of the complexity that can be allowed for possible attacks without causing any threat to the algorithm. In the evaluation process a number of different attacks that have been identified in the existing cryptanalytic literature were considered and lower bounds of the complexities of such attacks were determined. The Task Force concluded that the MILENAGE algorithm fulfils the requirements.

Another aspect of the security of the MILENAGE algorithm that needs consideration is the relationship between cryptographic and practical security requirements, some of which have already been given in Sections 8.2, 8.3.3 and 8.3.4. None of the requirements explicitly addresses the "cloning" problem. The main concern from the operator's point of view is to ensure that USIM cards are secure against cloning: first, it must be impossible to reproduce a valid USIM card and, second, it must be impossible, using existing USIM cards, to retrieve any information that would allow new valid cards to be produced.

The OP configuration field was required by the SA3, but its exact role in the configuration of security was left for the Task Force to define. The Task Force further analysed the role of OP, as presented in Section 8.3.4, where property 2 apparently addresses the problem of producing new, valid cards without authorization. The recommended implementation is to compute the secret OP_C parameter "off" the card using a strong one-way function in such a way that knowledge of the OP can never be retrieved from any feasible number of existing valid cards. Hence, the use of a secret OP configuration field, as long as it is never stored on the card, protects against unauthorized production of the operator's USIM cards.

The strength of the algorithm can only play a role if the memory of the card is protected against direct scanning of its contents. If this is the case, then the only possible way of retrieving subscriber-specific secret information from the card is to feed inputs to the card and analyse the corresponding outputs to piece together bits of secret information hidden inside the card's protected memory. In addition to input and output information, recent analysis methods make use of side channels to retrieve information on intermediate states as the card's processor executes the algorithm, and such side-channel information can be exploited for cryptanalysis of the algorithm.

According to property 3 in Section 8.3.4 one of OP's purposes is to keep certain algorithm details secret. What kinds of algorithm details can be kept secret? Recall that OP is a string of 128 bits. It is not possible to implement a large number of essentially different algorithms within each USIM card and use the secret OP to determine which of the algorithms is used. It might be possible to implement a small number of different algorithms on each USIM card and select the one that is used in a secret manner. But such an approach would multiply the complexity, which in the worst case would mean that each of the different algorithms would have to be weak to fit into the card. Hence, it is not surprising that the Task Force decided to come up with a single, strong algorithm and use the OP parameter as an additional key value to it. The main problem for cryptographers is how best to hide a relatively short secret value inside the cryptographic design in the most effective and secure manner. The best strategy cryptographers have come up with is to build the secret field of 128 bits that modifies and customizes an algorithm into the algorithm as a cryptographic key.

Consider the purely theoretical case where MILENAGE is totally broken and it becomes possible to retrieve subscriber secrets K and OP_C from the card in a systematic manner. Clearly it would not help to replace OP and generate new OP_C values for the subscribers, since they can be analysed using the same systematic manner. In such a case, the USIM cards would need to be replaced. Hence, the protection offered by MILENAGE against attacks that try to discover secret values from a subscriber's USIM card is solely based on its cryptographic strength. The Task Force's cryptanalysis team carried out extensive evaluations on the MILENAGE modes. The first results from voluntary evaluation efforts have recently appeared as well.

The main cryptographic strength of MILENAGE is due to its kernel block cipher. The chosen kernel algorithm (AES) has undergone extensive cryptanalysis by many different teams. Due to its status as a standard, its secure implementation and protection against side-channel attacks has also received much more attention than other algorithms. After Courtois and Pieprzyk published their analysis on AES [47], criticism of its security have increased, with many people believing its lifetime as a standard may be shorter than originally intended. Courtois and Pieprzyk's attack makes use of systems of overdefined algebraic equations, which the Rijndael block cipher was not designed to handle. Subsequently, a further alarming finding was made by Murphy and Robshaw [85], who were able to describe the entire relationships between the Rijndael plaintext, key and ciphertext by means of a sparse system of equations that involve at most quadratic (i.e., degree 1 or 2) polynomials over $GF(2^8)$. Opinion is divided as to the efficiency of these attacks, but some people estimate that the complexity of breaking the AES block cipher is about 2^{100} operations. Even if these vulnerabilities do not pose any practical threat, they do mean that the theoretical security of the kernel of MILENAGE has been significantly reduced and no longer meets the original requirement (Section 8.9.1).

Notation of Parameters, Sets and Functions

A	A register value in f8 and f9
AK	An anonymity key
AMF	An authentication management field
AND	The Boolean "and" operation
AUTN	An authentication token
AUTS	An authentication token in re-synchronization
AV	An authentication vector
B	A register value in f9
BEARER	The radio bearer identity
BLKCNT	The block counter in f8
BLOCK1	A keystream block for uplink encryption in GSM and ECSD
BLOCK2	A keystream block for downlink encryption in GSM and ECSD
C_1; C_2; C_3; C_4; C_5; C_6; C_7; C_8	Eight key modification constants in KASUMI
c_1; c_2; c_3; c_4; c_5	Five offset constants in MILENAGE specification
c1; c2; c3; c4; c5	Five conversion functions
C	Ciphertext
CIPHERTEXT	A ciphertext in f8
CA; CB; CC; CD; CE; CK	Bit string inputs to KGCORE
CK	A cipher (confidentiality) key in UMTS
CL	An integer input to KGCORE
CO	An output bitstream from KGCORE

COUNT	A frame-dependent input to A5/3
COUNT-C	A frame-dependent input to f8
COUNT-I	A frame-dependent input to f9
D	A data block in KASUMI specification
DIRECTION	A one-bit value to indicate whether encryption is uplink or downlink
$D \lll n$	The left circular rotation of data D by n bits
Δ	A limit value for SQN jump
ΔX_i	The difference between two data blocks
E_{K}	The encryption using algorithm E with key K
f_i	A round function (f-function) on round i
f_{K}	A function used in analysis of MISTY1
f, f^*	Functions used in the analysis of MILENAGE
f1; f2; f3; f4; f5	The UMTS AKA functions
f1*; f5*	The UMTS AKA functions in re-synchronization
FI	The inner function in KASUMI
FL	The key-dependent linear function in KASUMI
FO	The outer function in KASUMI
FRESH	A one-time random number chosen by RNC
Φ	A simplified counter mode construction
$\mathrm{GF}(2^7), \mathrm{GF}(2^8), \mathrm{GF}(2^9), \ldots, \mathrm{GF}(2^n)$	Finite fields (Galois fields)
GLC	The Global Counter
I	An input to f-function
i	An index
IBS	An input bit stream
IK	An integrity key
$\mathrm{IK}_{\mathrm{ESP}}$	An integrity key for ESP
IN1	An input to f1 and f1*
IND	An index value in some SQN
INPUT	A frame-dependent input to GEA3
IV; IV'	Initialization values
K	A subscriber authentication key in UMTS
K; K^*	Keys

K_j	A subkey (round key)
$\text{KASUMI}_{\text{CK}}$	The KASUMI encryption operation using key CK
$\text{KASUMI}_{\text{CK} \oplus \text{KM}}$	The KASUMI encryption operation using modified key CK $\oplus$ KM
$\text{KASUMI}_{\text{IK}}$	The KASUMI encryption operation using key IK
$\text{KASUMI}_{\text{IK} \oplus \text{KM}}$	The KASUMI encryption operation using modified key IK
Kc	A cipher key in GSM
KEYSTREAM	An ouput of the f8 keystream generator
KGORE	The general purpose keystream generator function
Ki	A subscriber authentication key in GSM
KI	A key used in FI
KL	A key used in FL
KLEN	The key length
KM	A key modifier
KO	A key used in FO
KSB	A keystream block in f8
KS[*i*]	The *i*th keystream bit in f8
L; L'; L_i	Left data halves
LENGTH	The length of plaintext in bits
M	The length of GEA keystream in octets
M; M'	Message inputs to a MAC function
MAC	A message authentication code (or function)
MAC-A	A message authentication code for network authentication
MAC-I	A message authentication code for integrity protection
MAC-S	A message authentication code for re-synchronization
MESSAGE	A message input to f9
n	An index
$n \bmod k$	The remainder when n is divided by k
NAND	The Boolean "negative-and" operation
NE-Id	A sending network element identifier
NULL	The identity function

O	An output of the f-function
OBS	An output bit stream
OP	An operator-dependent parameter in MILENAGE
OP_C	An operator and subscriber-dependent parameter in MILENAGE
OR	The Boolean "or" operation
Original component Id	A type of MAP message
OUTPUT	An output keystream from the GEA3 algorithm
OUTPUT$\{i\}$	The ith octet of the GEA3 keystream
OUT_1; OUT_2; OUT_3; OUT_4; OUT_5	Intermediate output values in MILENAGE specification
P; P_1; P_2	Plaintexts
PLAINTEXT	A plaintext in f8 specification
Prop	A proprietary field
PS	A padded string in f9
Q	An authentication quintuple
r_1; r_2; r_3; r_4; r_5	Five rotation constants in MILENAGE specification
R; R'; R_i	Right data halves
RES	An authentication response
RAND	A random (authentication) challenge
ROL	The left circular rotation of data by one bit
rot(X, r)	The rotation of data X by r bit positions toward the most significant bit
SEQ; SEQ1; SEQ2	Sequence numbers
SPI	A security parameter index
SRES	A signed response in GSM
SQN; SQN_{MS}	A sequence number; a sequence number suggested by mobile station
START	A starting value for the most significant part of the hyperframe number
SUM	The XOR sum of a number of data blocks
TEMP	An intermediate value in the MILENAGE specification
THRESHOLD	A maximal value for the START parameter

TR	The truncation operation in KASUMI specification
TVP	A time-variant parameter
u; u'; u''	Short notations for SQN ‖ AMF ‖ SQN ‖ AMF
v	An input to rot-function
w_i	A short notation for OUT_i, $i = 1, 2, 3, 4, 5$
W; W'	Prewhitening constants used in f8
W; W[i]	The output keystream of GEA (GPRS specification); the ith octet of W
X; X[i]	A data value; the ith bit of the data X
x; x'; x''	Short notations for RAND values
XMAC	An expected MAC value
XMAC-A	An expected MAC-A value
XMAC-I	An expected MAC-I value
XMAC-S	An expected MAC-S value
XOR	The Boolean "xor" operation
XRES	An expected RES value
y; y'; y''	Short notations for TEMP values
Z	The DIRECTION parameter to GEA
ZE	The zero-extension operation in KASUMI specification
$\|$	Concatenation of data strings
$\oplus$	The bit-wise XOR operation of data strings
$\cap$	The bit-wise AND operation of data strings
$\cup$	The bit-wise OR operation of data strings

Abbreviations

2G	Second Generation
3G	Third Generation
3GPP	Third Generation Partnership Project
3GPP2	Third Generation Partnership Project #2
A3	GSM authentication function
A5	GSM encryption algorithm
A5/1	GSM encryption algorithm #1
A5/2	GSM encryption algorithm #2
A5/3	GSM encryption algorithm #3
A8	GSM key generation function
AAA	Authorization, Authentication and Accounting
AAL	ATM Adaptation Layer
ACK	Acknowledge
ACL	Access Control List
ACO	Authentication Ciphering Offset
AES	Advanced Encryption Standard
AF	Authentication Framework
AH	Authentication Header
AK	Anonymity Key
AKA	Authentication and Key Agreement
AM	Acknowledged Mode
AMPS	Advanced Mobile Phone System
ANSI	American National Standards Institute
ARIB	Association of Radio Industries and Businesses
AS	Access Stratum
ASIC	Application Specific Integrated Circuit
ATM	Asynchronous Transfer Mode
AuC	Authentication Centre
BCH	Broadcast Channel
BER	Bit Error Rate

BMC	Broadcast/Multicast Control
BS	Base Station
BSS	Base Station Subsystem
BSS	Basic Service Set
BTS	Base Transceiver Station (GSM)
C-plane	Control plane
CA	Certification Authority
CBC	Cipher Block Chaining
CBC-MAC	Cipher Block Chaining Message Authentication Code
CC	Call Control
CCM	CBC-MAC
CCSA	China Communications Standards Association
CEPT	European Posts and Telecommunications Conference
CFN	Connection Frame Number
CFB	Cipher Feedback
CK	Cipher (Confidentiality) Key (UMTS)
CLPC	Closed Loop Power Control
CM	Communication Management
CN	Core Network
CPCH	Common Packet Channel
CRC	Cyclic Redundancy Check
CS	Circuit Switched
CSCF	Call Session Control Function
DCCH	Dedicated Control Channel
DCH	Dedicated Channel
DES	Data Encryption Standard
DIF	Difference
DMZ	Demilitarized Zone
DoI	Domain of Interpretation
DPA	Differential Power Analysis
DRNC	Drifting RNC
EAP	Extensible Authentication Protocol
ECB	Electronic Code Book
ECSD	Enhanced Circuit Switched Data
EDGE	Enhanced Data Rates for GSM Evolution
EGPRS	Enhanced GPRS
ESP	Encapsulation Security Payload
ESS	Extended Service Set
ETSI	European Telecommunications Institute
FACH	Forward Access Channel
FDD	Frequency Division Duplex
FDMA	Frequency Division Multiple Access

FIGS	Fraud Information Gathering System
FIPS	Federal Information Processing Standard
FM	Frequency Modulation
FPGA	Field Programmable Gate Array
FSE	Fast Software Encryption
FTP (ftp)	File Transfer Protocol
GEA	GPRS Encryption Algorithm
GERAN	GSM/EDGE Radio Access Network
GGSN	Gateway GPRS Support Node
GLC	Global Counter
GMSC	Gateway MSC
GPRS	General Packet Radio Service
GSM	Global System for Mobile Communications
GTP	GPRS Tunneling Protocol
HCI	Host Controller Interface
HE	Home Environment
HFN	Hyper Frame Number
HLR	Home Location Register
HMAC	(Keyed) Hashing for Message Authentication
HMAC-MD5	HMAC with MD5
HMAC-SHA-1	HMAC with SHA-1
HSS	Home Subscriber Server
HTTP	Hypertext Transfer Protocol
I-CSCF	Interrogating Call Session Control Function
IACR	International Association for Cryptologic Research
IBM	International Business Machines
IBSS	Independent Basic Service Set
ICC	Integrated Circuit Card
IDEA	International Data Encryption Algorithm
IEEE	Institute of Electrical and Electronics Engineers
IETF	Internet Engineering Task Force
IFF	Identification Friend or Foe
IK	Integrity Key
IKE	Internet Key Exchange
ILPC	Inner Loop Power Control
IMEI	International Mobile Equipment Identity
IMPI	IMS Private Identity
IMPU	IMS Public Identity
IMS	IP Multimedia CN Subsystem
IMSI	International Mobile Subscriber Identity
IMT-2000	International Mobile Telecommunications 2000
IP	Internet Protocol

IPR	Intellectual Property Right
IPsec	IP Security
ISIM	IMS Subscriber Identity Module
IST	Immediate Service Termination
IST	Information Society Technologies (European Union)
ITU	International Telecommunication Union
Iu	Interface between MSC/SGSN and RNC
Iub	Interface between RNC and BS
Iur	Interface between RNCs
KAC	Key Administration Centre
KSI	Key Set Identifier
LAI	Location Area Identity
LAN	Local Area Network
LCS	Location Services
LFSR	Linear Feedback Shift Register
LLC	Logical Link Control
MAC	Medium Access Control
MAC	Message Authentication Code
MAP	Mobile Application Part
MAPsec	MAP Security
MD5	Message Digest algorithm #5
ME	Mobile Equipment
MEA-1	MAP Encryption Algorithm #1
MExE	Mobile Execution Environment
MIME	Multipurpose Internet Mail Extension
MIT	Massachusetts Institute of Technology
MM	Mobility Management
MM1	Modified MISTY1
M^2	Modified MISTY
MS	Mobile Station
MSC	Mobile Switching Centre
NAI	Network Address Identifier
NAS	Non Access Stratum
NDS	Network Domain Security
NE	Network Element
NESSIE	New European Schemes for Signatures, Integrity and Encryption
NIST	National Institute of Standards
NMT	Nordic Mobile Telephone
NSA	National Security Agency
O&M	Operation and Management
OFB	Output Feedback

OLPC	Open Loop Power Control
OSI	Open Systems Interconnection
OTA	Over The Air
P-CSCF	Proxy Call Session Control Function
P-TMSI	Packet Temporary Mobile Subscriber Identity
PCH	Paging Channel
PDC	Personal (or Pacific) Digital Cellular
PDCP	Packet Data Convergence Protocol
PDP	Packet Data Protocol
PDU	Protocol Data Unit
PGP	Pretty Good Privacy
PHY	Physical Layer
PIN	Personal Identification Number
PKI	Public Key Infrastructure
PLMN	Public Land Mobile Network
PS	Packet Switched
PSK	Pre Shared Key
PSTN	Public Switched Telephone Network
QoS	Quality of Service
RACH	Random Access Channel
RADIUS	Remote Authentication Dial In User Service
RAI	Routing Area Identity
RAN	(Radio) Access Network
RANAP	Radio Access Network Application Protocol
RAT	Radio Access Technology
RC4	RSA security Cipher 4
RC5	RSA security Cipher 5
RFC	Request For Comments
RLC	Radio Link Control
RLC-SN	RLC Sequence Number
RNC	Radio Network Controller
RNTI	Radio Network Temporary Identity
RRC	Radio Resource Control
RRM	Radio Resource Management
RSN	Robust Security Network
S-CSCF	Serving Call Session Control Function
S-MIME	Secured MIME
SA	Security Association
SA3	Services and System Aspects Working Group 3 (Security)
SAD	Security Association Database
SAGE TF 3GPP	SAGE Task Force for 3GPP Algorithms
SCCP	Signalling Connection Control Part

SDP	Session Description Protocol
SDU	Signalling Data Unit
SGSN	Serving GPRS Support Node
SHA-1	Secure Hash Algorithm #1
SIG	(Bluetooth) Special Interest Group
SIM	Subscriber Identity Module
SIP	Session Initiation Protocol
SIR	Signal-to-Interface Ratio
SM	Session Management
SMSC	Short Message Service Centre
SN	Serving Network
SPA	Simple Power Analysis
SPI	Security Parameter Index
SPN	Substitution Permutation Network
SQN	A sequence number
SQN_{AuC}	A sequence number suggested by AuC
SQN_{MS}; SQN_{USIM}	A sequence number suggested by MS, or more specifically, by USIM
SRNC	Serving RNC
SS	Supplementary Service
SS7	Signalling System #7
SSID	Service Set Identity
SSL	Secure Socket Layer
T1 (ANSI T1)	Standards Committee on Telecommunications
TCP	Transmission Control Protocol
TDD	Time Division Duplex
TDMA	Time Division Multiple Access
tel URL	URL for telephone calls
TF	Task Force
TIA	Telecommunications Industry Association
TKIP	Temporal Key Integrity Protocol
TLS	Transport Layer Security
TMSI	Temporary Mobile Subscriber Identity
TR	Technical Report (ETSI)
TS	Technical Specification (ETSI)
TTA	Telecommunications Technology Association
TTC	Telecommunications Technology Council
TTI	Transmission Time Interval
U-Plane	User Plane
UA	User Agent
UAC	User Agent Client
UAS	User Agent Server

UDP	User Datagram Protocol
UE	User Equipment
UEA	UMTS Encryption Algorithm
UIA1	UMTS Integrity Algorithm 1
UICC	Universal Integrated Circuit Card
UM	Unacknowledged Mode
UMTS	Universal Mobile Telecommunications System
URI	Uniform Resource Identifier
URL	Uniform Resource Locator
USIM	Universal Subscriber Identity Module
UTC	Co-ordinated Universal Time
UTRAN	UMTS Terrestrial Radio Access Network
Uu	Radio interface
VLR	Visitor Location Register
VPN	Virtual Private Network
WAP	Wireless Application Protocol
WCDMA	Wideband Code Division Multiple Access
WEP	Wired Equivalent Privacy
Wi-Fi	Wireless Fidelity
WLAN	Wireless LAN
WPA	Wi-Fi Protected Access
WRAP	Wireless Robust Authentication Protocol
Zd	Interface between KACs of different PLMNs
Ze	Interface between KAC and NE within a PLMN
Zf	Interface between NEs of different PLMNs

References

[1] 3GPP TS 23.057 V5.1.0 (2002-9) Technical Specification; Third Generation Partnership Project; Technical Specification Group Terminals; Mobile Execution Environment (MExE); Functional description; Stage 2 (Release 5).

[2] 3GPP TS 23.060 V5.4.0 (2002-12) Technical Specification; Third Generation Partnership Project; Technical Specification Group Services and System Aspects; General Packet Radio Service (GPRS); Service description; Stage 2 (Release 5).

[3] 3GPP TS 23.228 V5.7.0 (2002-12) Technical Specification; Third Generation Partnership Project; Technical Specification Group Services and System Aspects; IP Multimedia Subsystem (IMS); Stage 2 (Release 5).

[4] 3GPP TS 24.228 V5.3.0 (2002-12) Technical Specification; Third Generation Partnership Project; Technical Specification Group Core Network; Signalling flows for the IP multimedia call control based on SIP and SDP; Stage 3 (Release 5).

[5] 3GPP TS 24.229 V5.3.0 (2002-12) Technical Specification; Third Generation Partnership Project; Technical Specification Group Core Network; IP multimedia call control protocol based on Session Initiation Protocol (SIP) and Session Description Protocol (SDP); Stage 3 (Release 5).

[6] 3GPP TS 25.212 V5.3.0 (2002-12) Technical Specification; Third Generation Partnership Project; Technical Specification Group Radio Access Network; Multiplexing and channel coding (FDD) (Release 5).

[7] 3GPP TS 29.002 V5.4.0 (2002-12) Technical Specification; Third Generation Partnership Project; Technical Specification Group Core Network; Mobile Application Part (MAP) specification; (Release 5).

[8] 3GPP TR 31.900 V5.1.0 (2002-6) Technical Report; Third Generation Partnership Project; Technical Specification Group Terminals; SIM/USIM internal and external interworking aspects (Release 5).

[9] 3GPP TS 33.102 V5.2.0 (2003-06) Technical Specification; Third Generation Partnership Project; Technical Specification Group Services and System Aspects; 3G Security; Security architecture (Release 5).

[10] 3GPP TS 33.105 V4.1.0 (2001-06) Technical Specification; Third Generation Partnership Project; Technical Specification Group Services and System Aspects; 3G Security; Cryptographic algorithm requirements (Release 4).

[11] 3GPP TS 33.120 V4.0.0 (2001-03) Technical Specification; Third Generation Partnership Project; Technical Specification Group Services and System Aspects; 3G Security; Security principles and objectives (Release 4).

[12] 3GPP TS 33.200 V5.1.0 (2002-12) Technical Specification; Third Generation Partnership Project; Technical Specification Group Services and System Aspects; 3G Security; Network Domain Security; MAP application layer security (Release 5).

[13] 3GPP TS 33.210 V5.4.0 (2003-06) Technical Specification; Third Generation Partnership Project; Technical Specification Group Services and System Aspects; 3G Security; Network Domain Security; IP network layer security (Release 5).

[14] 3G TR 33.900 V1.2.0 (2000-01) Technical Specification; Third Generation Partnership Project; Technical Specification Group SA WG3; A guide to Third Generation security (3G TR 33.900 version 1.2.0).

[15] 3GPP TR 33.901 V4.0.0 (2001-09) Technical Report; Third Generation Partnership Project; Technical Specification Group Services and System Aspects; 3G Security; Criteria for cryptographic algorithm design process (Release 4).

[16] 3GPP TR 33.908 V4.0.0 (2001-09) Technical Report; Third Generation Partnership Project; Technical Specification Group Services and System Aspects; 3G Security; General report on the design, specification [*sic*] and evaluation of 3GPP standard confidentiality and integrity algorithms (Release 4).

[17] 3GPP TR 33.909 V1.0.0 (2000-12) Technical Report; Third Generation Partnership Project; Technical Specification Group Services and System Aspects; Report on the evaluation of 3GPP standard confidentiality and integrity algorithms (Release 1999).

[18] 3GPP TR 33.909 V4.0.1 (2001-06) Technical Report; Third Generation Partnership Project; Technical Specification Group Services and System Aspects; 3G Security; Report on the Design and Evaluation of the MILENAGE Algorithm Set; Deliverable 5: An example algorithm for the 3GPP authentication and key generation functions (Release 4) [the body of this document is identical to that of TS 35.909].

[19] 3GPP TS 35.201 V5.0.0 (2002-06) Technical Specification; Third Generation Partnership Project; Technical Specification Group Services and System Aspects; 3G Security; Specification of the 3GPP Confidentiality and Integrity Algorithms; Document 1: f8 and f9 specification (Release 5).

[20] 3GPP TS 35.202 V5.0.0 (2002-06) Technical Specification; Third Generation Partnership Project; Technical Specification Group Services and System Aspects; 3G Security; Specification of the 3GPP Confidentiality and Integrity Algorithms; Document 2: KASUMI specification (Release 5).

[21] 3GPP TS 35.203 V5.0.0 (2002-06) Technical Specification; Third Generation Partnership Project; Technical Specification Group Services and System Aspects; 3G Security; Specification of the 3GPP Confidentiality and Integrity Algorithms; Document 3: Implementors' test data (Release 5).

[22] 3GPP TS 35.204 V5.0.0 (2002-06) Technical Specification; Third Generation Partnership Project; Technical Specification Group Services and System Aspects; 3G Security; Specification of the 3GPP Confidentiality and Integrity Algorithms; Document 4: Design conformance test data (Release 5).

[23] 3GPP TS 35.205 V5.0.0 (2002-06) Technical Specification; Third Generation Partnership Project; Technical Specification Group Services and System Aspects; 3G Security; Specification of the MILENAGE Algorithm Set: An example algorithm set for the 3GPP authentication and key generation functions f1, f1*, f2, f3, f4, f5 and f5*; Document 1: General (Release 5).

[24] 3GPP TS 35.206 V5.1.0 (2003-06) Technical Specification; Third Generation Partnership Project; Technical Specification Group Services and System Aspects; 3G Security; Specification of the MILENAGE Algorithm Set: An example algorithm set for the 3GPP authentication and key generation functions f1, f1*, f2, f3, f4, f5 and f5*; Document 2: Algorithm specification (Release 5).

[25] 3GPP TS 35.207 V5.0.0 (2002-06) Technical Specification; Third Generation Partnership Project; Technical Specification Group Services and System Aspects; 3G Security; Specification of the MILENAGE Algorithm Set: An example algorithm set for the 3GPP authentication and key generation functions f1, f1*, f2, f3, f4, f5 and f5*; Document 3: Implementors' test data (Release 5).

[26] 3GPP TS 35.208 V5.0.0 (2002-06) Technical Specification; Third Generation Partnership Project; Technical Specification Group Services and System Aspects; 3G Security; Specification of the MILENAGE Algorithm Set: An example algorithm set for the 3GPP authentication and key generation functions f1, f1*, f2, f3, f4, f5 and f5*; Document 4: Design conformance test data (Release 5).

[27] 3GPP TR 35.909 V5.0.0 (2002-05) Technical Report; Third Generation Partnership Project; Technical Specification Group Services and System Aspects; 3G Security; Specification of the MILENAGE Algorithm Set: An example algorithm set for the 3GPP authentication and key generation functions f1, f1*, f2, f3, f4, f5 and f5*; Document 5: Summary and results of design and evaluation (Release 5).

[28] 3GPP TS 41.061 V4.0.0 (2001-03) Technical Specification; Third Generation Partnership Project; Technical Specification Group Services and System Aspects; Digital Cellular Telecommunications System (Phase 2+); General Packet Radio Service (GPRS); GPRS ciphering algorithm requirements (Release 4).

[29] 3GPP TS 43.020 V5.0.0 (2002-07) Technical Specification; Third Generation Partnership Project; Technical Specification Group Services and System Aspects; Security-related network functions (Release 5).

[30] 3GPP TS 55.205 V6.0.0 (2002-12) Technical Specification; Third Generation Partnership Project; Technical Specification Group Services and System Aspects; Specification of the GSM-MILENAGE Algorithms: An example algorithm set for the GSM authentication and key generation functions A3 and A8 (Release 6).

[31] 3GPP TS 55.216 V6.1.0 (2002-12) Technical Specification; Third Generation Partnership Project; Technical Specification Group Services and System Aspects; 3G Security; Specification of the A5/3 Encryption Algorithms for GSM and ECSD, and the GEA3 Encryption Algorithm for GPRS; Document 1: A5/3 and GEA3 specifications (Release 6).

[32] 3GPP TS 55.217 V6.1.0 (2002-12) Technical Specification; Third Generation Partnership Project; Technical Specification Group Services and System Aspects; 3G Security; Specification of the A5/3 Encryption Algorithms for GSM and ECSD, and the GEA3 Encryption Algorithm for GPRS; Document 2: Implementors' test data (Release 6).

[33] 3GPP TS 55.218 V6.1.0 (2002-12) Technical Specification; Third Generation Partnership Project; Technical Specification Group Services and System Aspects; 3G Security; Specification of the A5/3 Encryption Algorithms for GSM and ECSD, and the GEA3 Encryption Algorithm for GPRS; Document 3: Design conformance test data (Release 6).

[34] 3GPP TR 55.919 V6.1.0 (2002-12) Technical Report; Third Generation Partnership Project; Technical Specification Group Services and System Aspects; 3G Security; Specification of the A5/3 Encryption Algorithms for GSM and ECSD, and the

GEA3 Encryption Algorithm for GPRS; Document 4: Design and evaluation report (Release 6).

[35] J. Arkko and R. Blom, The MAP Security Domain of Interpretation for Internet Security Association and Key Management Protocol (Internet draft, 27 May 2002, work in progress), http://www.ietf.org/internet-drafts/draft-arkko-map-doi-07.txt

[36] S. Babbage, Design of Security Algorithms for Third Generation Mobile Telephony. *Information Security Technical Report*, **5**(3), 2000, 66–73.

[37] S. Babbage and L. Frisch, On MISTY1 higher order differential cryptanalysis, in: *Proceedings of ICISC 2000* (Lecture Notes in Computer Science No. 2015), Springer-Verlag 2000, 22–36.

[38] M. Bellare, J. Kilian and P. Rogaway, The security of cipher block chaining message authentication code, *Journal of Computer and System Sciences*, **61**(3), December 2000, 362–399. Previous version published in: Yvo Desmedt (ed.), *Advances in Cryptology—Crypto '94* (Lecture Notes in Computer Science No. 839), Springer-Verlag 1994, 341–358. Full paper available at http://charlotte.ucsd.edu/users/mihir/papers/cbc.html

[39] M. Bellare, A. Desai, E. Jokipii and P. Rogaway, A Concrete Security Treatment of Symmetric Encryption: Analysis of the DES Modes of Operation, *Proceedings of 38th Annual Symposium on Foundations of Computer Science, IEEE, 1997.* Full paper available at http://charlotte.ucsd.edu/users/mihir/papers/sym-enc.html

[40] T. Beth and C. Ding, On almost perfect nonlinear permutations, in: T. Helleseth (ed.), *Advances in Cryptology—Eurocrypt '93* (Lecture Notes in Computer Science No. 765), Springer-Verlag, 1994, 65–76.

[41] E. Biham and A. Shamir, Differential cryptanalysis of DES-like cryptosystems, in: A. J. Menezes and S. A. Vanstone (eds), *Advances in Cryptology—Crypto '90* (Lecture Notes in Computer Science No. 537), Springer-Verlag, 1991, 2–21.

[42] Bluetooth SIG, *Specification of the Bluetooth System* (Vol. 1, Version 1.1), February 2001.

[43] M. Briceno, I. Goldberg and D. Wagner, GSM cloning, 1998, http://www.isaac.cs.berkeley.edu/isaac/gsm-faq.html

[44] M. Burmester, On the risk of opening distributed keys. In: Y. Desmedt (ed.), *Advances in Cryptology—Crypto '94* (Lecture Notes in Computer Science No. 839), Springer-Verlag, 1994, 308–317.

[45] A. Canteaut and M. Videau, Degree of composition of highly nonlinear functions and applications to higher order differential cryptanalysis, in: L. Knudsen (ed.), *Advances in Cryptology—Eurocrypt 2002* (Lecture Notes in Computer Science No. 2332), Springer-Verlag, 2002, 518–533.

[46] D. Coppersmith, The Data Encryption Standard (DES) and its strength against attacks. *IBM Technical Journal on Research and Development*, **38**, 1994, 243–250.

[47] N. Courtois and J. Pieprzyk, Cryptanalysis of block ciphers with overdefined systems of equations, *Advances in Cryptology—Asiacrypt 2002* (Lecture Notes in Computer Science No. 2501), Springer-Verlag, 2002, 267–287.

[48] D. Joan, Correlation matrices, in: B. Preneel (ed.), *Proceedings of Fast Software Encryption '94 Conference* (Lecture Notes in Computer Science No. 1008), 1995, 275–285.

[49] W. Diffie, The first ten years of public key cryptology. *Proceedings of the IEEE*, **76**, 1988, 560–577.

[50] H. Dobbertin, Almost perfect nonlinear power functions on GF(2^n): The Welch case, *IEEE Transactions on Information Theory*, **45**(4), May 1999.
[51] ETSI EG 202 238 V1.1.1 (2003-08) ETSI Guide; TIPHON™; Evaluation criteria for cryptographic algorithms; DEG/TIPHON-08007.
[52] H. Gaines, *Cryptanalysis: A Study of Ciphers and Their Solutions*, Dover Publications, New York, 1956.
[53] H. Gilbert, The Security of "one-block-to-many" modes of operation, in: T. Johansson (ed.), *Fast Software Encryption—FSE 2003* (Lecture Notes in Computer Science), Springer-Verlag (to appear).
[54] H. Gilbert and M. Minier, New results on the pseudorandomness of some block cipher constructions, in: M. Matsui (ed.) *Fast Software Encryption—FSE 2001* (Lecture Notes in Computer Science No. 2355), Springer-Verlag, 2002, 248–266.
[55] T. Helleseth, C. Rong and D. Sandberg, New families of almost perfect nonlinear power mappings, *IEEE Transactions Information Theory*, **45**, 1999, 475–485.
[56] H. Holma and A. Toskala (eds), *WCDMA for UMTS: Radio Access for Third Generation Mobile Communications*, John Wiley & Sons, Chichester, UK, 2002.
[57] D. W. Hong, J. S. Kang, B. Preneel and H. Ryu, The security of 3GPP-MAC on the reduction-based cryptography, in: T. Johansson (ed.), *Fast Software Encryption—FSE 2003* (Lecture Notes in Computer Science), Springer-Verlag (to appear).
[58] IEEE Std 802.11 (1999 edition) Part 11: Wireless LAN Medium Access Control (MAC) and Physical Layer (PHY) Specifications.
[59] IEEE Std 802.1X (2001) Standards for Local and Metropolitan Area Networks: Port Based Access Control, June 2001.
[60] IEEE Std 802.11i/D3.02, Draft Supplement to ISO/IEC 8802-11/1999(I) ANSI/IEEE Std 802.11 (1999 edition) Specification for Robust Security, April 2003.
[61] ISO/IEC 9797-1 (1999) Information Technology; Security Techniques; Message Authentication Codes (MACs); Part 1: Mechanisms using a block cipher.
[62] ISO/IEC 9797-2 (2002) Information Technology; Security Techniques; Message Authentication Codes (MACs); Part 2: Mechanisms using a dedicated hash function.
[63] ISO/IEC 9798-4 (1999) Information Technology; Security Techniques; Entity Authentication; Part 4: Mechanisms using a cryptographic check function.
[64] T. Jakobsen and L. R. Knudsen, Attacks on block ciphers of low algebraic degree, *Journal of Cryptology*, **14**(3), 2001, 197–210.
[65] H. Kaaranen, A. Ahtiainen, L. Laitinen, S. Naghian and V. Niemi, *UMTS Networks: Architecture, Mobility and Services*, John Wiley & Sons, Chichester, UK, 2001.
[66] J. S. Kang, O. Y. Yi, D. W. Hong and H. S. Cho, Pseudorandomness of MISTY-type transformations and the block cipher KASUMI, *ACISP 2001* (Lecture Notes in Computer Science No. 2119), Springer-Verlag, 2001, 60–73.
[67] J. S. Kang, S. U. Shin, D. W. Hong and O. Y. Yi, Provable security of KASUMI and 3GPP encryption mode f8, *Advances in Cryptology—Asiacrypt 2001* (Lecture Notes in Computer Science No. 2248), Springer-Verlag, 2001.
[68] L. R. Knudsen, *DEAL—A 128-bit Block Cipher* (Technical Report #151), University of Bergen, Department of Informatics, Norway, February 1998. Submitted as a candidate for the Advanced Encryption Standard. Part of the NIST public record of the development of the Advanced Encryption Standard, see `http://www.nist.gov/aes`.
[69] L. R. Knudsen, Contemporary block ciphers, in: I. Damgård (ed.), *Lectures on Data Security* (Lecture Notes in Computer Science Tutorial), Springer-Verlag, 1999, 103–126.

[70] L. R. Knudsen and C. J. Mitchell, An analysis of the 3GPP-MAC scheme, in: D. Augot and C. Carlet (eds), *Workshop on Coding and Cryptography, WCC 2001*, Les Ecoles de Coëtquidan, 2001, 319–328.

[71] L. R. Knudsen and D. Wagner, Integral cryptanalysis, in: J. Daemen and V. Rijmen (eds), *Fast Software Encryption—FSE 2002* (Lecture Notes in Computer Science No. 2365), Springer-Verlag, 2002, 112–127.

[72] G. Kolata, When science inadvertently aids an enemy, *New York Times*, 25 September 2001.

[73] M. Krause, BDD-based cryptanalysis of keystream generators, in: *Eurocrypt 2002 Proceedings*, Springer-Verlag, 2002.

[74] U. Kühn, Cryptanalysis of reduced-round MISTY, in: B. Pfitzmann (ed.), *Advances in Cryptology—Eurocrypt 2001* (Lecture Notes in Computer Science No. 2045), Springer-Verlag, 2001, 325–339.

[75] U. Kühn, Improved cryptanalysis of MISTY1, in: J. Daemen and V. Rijmen (eds), *Fast Software Encryption—FSE 2002* (Lecture Notes in Computer Science No. 2365), Springer-Verlag, 2002, 61–75.

[76] X. Lai, J. Massey and S. Murphy, Markov ciphers and differential cryptanalysis, in: D. Davies (ed.), *Advances in Cryptology—Eurocrypt '91* (Lecture Notes in Computer Science No. 547), Springer-Verlag, 1991, 17–38.

[77] X. Lai, Higher order derivatives and differential cryptanalysis, in: *Proceedings of Symposium on Communication, Coding and Cryptography* (in honour of James L. Massey on the occasion of his 60th birthday), *10–13 February 1994, Monte Verità, Ascona, Switzerland.*

[78] S. Landau, Communications security for the twenty-first century: the Advanced Encryption Standard. *Notices of the AMS*, **47**, 2000, 450–459.

[79] L. Marks, *Between Silk and Cyanide*, HarperCollins, 2000.

[80] J. Massey, An introduction to contemporary cryptology, *Proceedings of the IEEE*, **76**, 1988, 533–549.

[81] M. Matsui, Linear cryptanalysis of DES, in: T. Helleseth (ed.), *Advances in Cryptology—Eurocrypt '93* (Lecture Notes in Computer Science No. 765), Springer-Verlag, 1994, 386–397.

[82] M. Matsui, New structure of block ciphers with provable security against differential and linear cryptanalysis, in: D. Gollman (ed.), *Fast Software Encryption—FSE '96* (Lecture Notes in Computer Science No. 1039), Springer-Verlag, 1996, 205–218.

[83] M. Matsui, New block encryption algorithm MISTY, in E. Biham (ed.), *Fast Software Encryption—FSE '97* (Lecture Notes in Computer Science No. 1267), Springer-Verlag, 54–68.

[84] A. Menezes, P. C. van Oorschot and S. A. Vanstone, *Handbook of Applied Cryptography*, CRC Press, Boca Raton, FL, 1997.

[85] S. Murphy and M. Robshaw, Essential algebraic structure within the AES, in: M. Yung (ed.), *Advances in Cryptology—CRYPTO 2002* (Lecture Notes in Computer Science No. 2442), Springer-Verlag, 2002, 1–16.

[86] NESSIE Project home page:
`https://www.cosic.esat.kuleuven.ac.be/nessie/`

[87] NESSIE, NESSIE security report, Deliverable 20, Version 1.0, 21 October 2002.

[88] NIST AES home page: `http://csrc.nist.gov/CryptoToolkit/aes/`

[89] NIST FIPS PUB 81, DES Modes of Operation,
`http://csrc.nist.gov/publications/fips/fips81/fips81.htm`

[90] NIST FIPS PUB 197, Announcing the Advanced Encryption Standard, Specification for the Advanced Encryption Standard (AES), 26 November 2001, http://csrc.nist.gov/publications/fips/fips197/fips-197.pdf

[91] NIST Report; James Nechvatal, Elaine Barker, Lawrence Bassham, William Burr, Morris Dworkin, James Foti and Edward Roback, Report on the Development of the Advanced Encryption Standard (AES), 2 October 2000, http://csrc.nist.gov/CryptoToolkit/aes/round2/r2report.pdf

[92] NIST Special Publication 800-38A; Morris Dworkin, Recommendation for Block Cipher Modes of Operation, December 2001, http://csrc.nist.gov/publications/nistpubs/800-38a/sp800-38a.pdf

[93] K. Nyberg, Differentially uniform mappings for cryptography, in: T. Helleseth (ed.), *Advances in Cryptology—Eurocrypt '93* (Lecture Notes in Computer Science No. 765), Springer-Verlag, 1994, 55–64.

[94] K. Nyberg, Linear approximation of block cipher, in: A. De Santis (ed.), *Advances in Cryptology—Eurocrypt '94* (Lecture Notes in Computer Science No. 950), Springer-Verlag, 1995, 439–444.

[95] K. Nyberg and L. Knudsen, Provable security against a differential attack. *Journal of Cryptology*, **8**(1), 1995.

[96] T. Rappaport, *Wireless Communications*, Prentice Hall, New Jersey, 2002.

[97] RFC 2045, Multipurpose Internet Mail Extensions (MIME), Part One: Format of Internet Message Bodies. N. Freed, N. Borenstein. November 1996.

[98] RFC 2104, HMAC: Keyed-Hashing for Message Authentication. H. Krawczyk, M.Bellare, and R. Canetti. February 1997.

[99] RFC 2138, Remote Authentication Dial-In User Service (RADIUS). C. Rigney, A. Rubens, W. Simpson and S. Willens. April 1997.

[100] RFC 2246, The TLS Protocol Version 1.0. T. Dierks and C. Allen. January 1999.

[101] RFC 2284, PPP Extensible Authentication Protocol (EAP). L. Blunk and J. Vollbrecht. March 1998.

[102] RFC 2327, SDP: Session Description Protocol. M. Handley and V. Jacobson. April 1998.

[103] RFC 2401, Security Architecture for the Internet Protocol. S. Kent and R. Atkinson. November 1998.

[104] RFC 2402, IP Authentication Header. S. Kent and R. Atkinson. November 1998.

[105] RFC 2403, The Use of HMAC-MD5-96 within ESP and AH. C. Madson and R. Glenn. November 1998.

[106] RFC 2404, The Use of HMAC-SHA-1-96 within ESP and AH. C. Madson and R. Glenn. November 1998.

[107] RFC 2405, The ESP DES-CBC Cipher Algorithm With Explicit IV. C. Madson and N. Doraswamy. November 1998.

[108] RFC 2406, IP Encapsulating Security Payload (ESP). S. Kent and R. Atkinson. November 1998.

[109] RFC 2407, The Internet IP Security Domain of Interpretation for ISAKMP. D. Piper. November 1998.

[110] RFC 2408, Internet Security Association and Key Management Protocol (ISAKMP). D. Maughan, M. Schertler, M. Schneider and J. Turner. November 1998.

[111] RFC 2409, The Internet Key Exchange (IKE). D. Harkins and D. Carrel. November 1998.

[112] RFC 2410, The NULL Encryption Algorithm and Its Use With IPsec. R. Glenn and S. Kent. November 1998.

[113] RFC 2411, IP Security Document Roadmap. R. Thayer, N. Doraswamy and R. Glenn. November 1998.

[114] RFC 2412, The OAKLEY Key Determination Protocol. H. Orman. November 1998.

[115] RFC 2486, The Network Access Identifier. B. Aboba and M. Beadles. January 1999.

[116] RFC 2616, Hypertext Transfer Protocol—HTTP/1.1. R. Fielding, J. Gettys, J. Mogul, H. Frystyk, L. Masinter, P. Leach and T. Barners-Lee. June 1999.

[117] RFC 2617, HTTP Authentication: Basic and Digest Access Authentication. J. Franks, P. Hallam-Baker, J. Hostetler, S. Lawrence, P. Leach, A. Luotonen and L. Stewart. June 1999.

[118] RFC 2806, URLs for Telephone Calls. A. Vähä-Sipilä. April 2000.

[119] RFC 2976, The SIP INFO Method. S. Donovan. October 2000.

[120] RFC 3261, SIP: Session Initiation Protocol. J. Rosenberg, H. Schulzrinne, G. Camarillo, A. Johnston, J. Peterson, R. Sparks, M. Handley and E. Schooler. June 2002.

[121] RFC 3266, Support for IPv6 in Session Description Protocol (SDP). S. Olson, G. Camarillo and A. B. Roach. June 2002.

[122] RFC 3310, Hypertext Transfer Protocol (HTTP) Digest Authentication Using Authentication and Key Agreement (AKA). A. Niemi, J. Arkko and V. Torvinen. September 2002.

[123] RFC 3329, Security Mechanism Agreement for the Session Initiation Protocol (SIP). J. Arkko, V. Torvinen, G. Camarillo, A. Niemi and T. Haukka. January 2003.

[124] RFC 3428, Session Initiation Protocol (SIP) Extension for Instant Messaging. B. Campbell, ed., J. Rosenberg, H. Schulzrinne, C. Huitema and D. Gurle. December 2002.

[125] G. Roelofsen, Cryptographic algorithms in telecommunications systems, *Information Security Technical Report*, **4**(1), (1999), 29–37.

[126] O. S. Rothaus, On "bent" functions, *Journal of Combinatorial Theory, Series A*, **20**, 1976, 300–305.

[127] R. A. Rueppel, Stream ciphers, in: G. Simmons (ed.), *Contemporary Cryptology*, IEEE Press, 1992, 65–124.

[128] K. Sakurai and Y. Zheng, On non-pseudorandomness of block ciphers with provable immunity against linear cryptanalysis, *IEICE Trans. Fundamentals*, **E80-A**(1), 1997, 19–24.

[129] C. Shannon, A mathematical theory of communication, *Bell Systems Technical Journal*, **27**, 1948, 379–423, 623–656.

[130] C. Shannon, Communication theory of secrecy systems, *Bell Systems Technical Journal*, **28**, 1949, 656–715.

[131] D. Stinson, *Cryptography, Theory and Practise* (2nd edn), Chapman & Hall/CRC Press, London/Boca Raton, FL, 2002.

[132] M. Sugita, *Higher Order Differential Attack on Block Cipher MISTY1, 2* (Technical Report ISEC98-4), Institute of Electronics, Information and Communication Engineers (IEICE), 1998.

[133] H. Tanaka, Kazuyuki Hisamatsu and Toshinobu Kaneko, Strength of MISTY1 without FL function for higher order differential attack, in: *Applied Algebra, Algebraic Algorithms and Error Correcting Codes* (Lecture Notes in Computer Science No. 1719), Springer-Verlag, 1999, 221–230.

[134] Wassenaar Arrangement, December 2002, `http://www.wassenaar.org/`

[135] M. V. Wilkes, *Time-Sharing Computer Systems* (2nd edn), American Elsevier, New York, 1972.

[136] 3GPP TS 33.203 V5.6.0 (2003-6) Technical Specification; Third Generation Partnership Project; Technical Specification Group Services and System Aspects; 3G Security; Access security for IP-based services (Release 5).

Index